The Midlife Crisis of the Nuclear Nonproliferation Treaty

The Midlife Crisis of the Nuclear Nonproliferation Treaty

Peter Pella
Professor of Physics Emeritus, Gettysburg College, PA, USA

Morgan & Claypool Publishers

Copyright © 2016 Morgan & Claypool Publishers

All rights reserved. No part of this publication may be reproduced, stored in a retrieval system or transmitted in any form or by any means, electronic, mechanical, photocopying, recording or otherwise, without the prior permission of the publisher, or as expressly permitted by law or under terms agreed with the appropriate rights organization. Multiple copying is permitted in accordance with the terms of licences issued by the Copyright Licensing Agency, the Copyright Clearance Centre and other reproduction rights organisations.

Rights & Permissions
To obtain permission to re-use copyrighted material from Morgan & Claypool Publishers, please contact info@morganclaypool.com.

ISBN   978-1-6817-4389-9 (ebook)
ISBN   978-1-6817-4388-2 (print)
ISBN   978-1-6817-4391-2 (mobi)

DOI   10.1088/978-1-6817-4389-9

Version: 20160301

IOP Concise Physics
ISSN 2053-2571 (online)
ISSN 2054-7307 (print)

A Morgan & Claypool publication as part of IOP Concise Physics
Published by Morgan & Claypool Publishers, 40 Oak Drive, San Rafael, CA, 94903, USA

IOP Publishing, Temple Circus, Temple Way, Bristol BS1 6HG, UK

To Eleanor for her unending love and support.

Contents

Preface		ix
Acknowledgments		x
Author biography		xi

1	Introduction	1-1

2	Scientific background	2-1
2.1	Fundamental forces	2-1
2.2	Model of the atom	2-2
2.3	Isotopes	2-4
2.4	Stability	2-5
2.5	Half-life	2-5
2.6	Fission and fusion	2-5

3	Technology	3-1
3.1	Reactor physics	3-1
3.2	Nuclear reactors	3-3
3.3	Reactor designs	3-5
3.4	The nuclear fuel cycle and nuclear proliferation	3-8
3.5	Uranium enrichment	3-8
3.6	Reactor wastes	3-9
3.7	Reprocessing	3-10
3.8	Fission weapons	3-10
3.9	Gun-barrel device	3-11
3.10	Implosion device	3-12
3.11	Proliferation concerns	3-13
	References	3-14

4	The nuclear nonproliferation treaty	4-1
4.1	Why proliferation is a concern	4-2
4.2	NPT early history	4-3
4.3	Conference on Disarmament	4-4
4.4	The NPT	4-4
4.5	Recent history of the NPT	4-7

4.6	The IAEA, the Nuclear Suppliers Group and the Zangger Committee	4-12
	4.6.1 The IAEA	4-12
	4.6.2 The NSG	4-14
	4.6.3 Zangger Committee	4-15
4.7	Successes of the NPT	4-15
4.8	Difficulties faced by the NPT	4-16
	4.8.1 Iraq	4-16
	4.8.2 Libya	4-17
	4.8.3 Syria	4-17
	4.8.4 DPRK	4-18
	4.8.5 Iran	4-19
4.9	Lessons learned	4-21
	References	4-22
5	**The NPT in crisis**	**5-1**
5.1	Safeguards	5-1
5.2	Technical cooperation	5-2
5.3	Disarmament	5-5
5.4	Middle East	5-9
5.5	Security assurances	5-10
	References	5-12
6	**Conclusion**	**6-1**
	Reference	6-3
Appendix	**The NPT Treaty**	**7-1**

Preface

The Nuclear Nonproliferation Treaty (NPT) has been the principle legal barrier to prevent the spread of nuclear weapons for the past forty-five years. It promotes the peaceful uses of nuclear technology and insures, through the application of safeguards inspections conducted by the International Atomic Energy Agency (IAEA), that those technologies are not being diverted toward the production of nuclear weapons. It is also the only multinational treaty that obligates the five nuclear weapons states that are party to the treaty (China, France, Great Britain, Russia, and the United States) to pursue nuclear disarmament measures.

Though there have been many challenges over the years, most would agree that the treaty has largely been successful. However many are concerned about the continued viability of the NPT. The perceived slow pace of nuclear disarmament, the interest by some countries to consider a weapons program while party to the treaty, and the funding and staffing issues at the IAEA, are all putting considerable strain on the treaty. This book explores those issues and offers some possible solutions to insure that the NPT will survive effectively for many years to come.

Acknowledgments

There have been many people who have made significant contributions in my life who deserve to be remembered. I want to thank John Winhold and John Watson for instilling in me a love for Physics, and Hendrix College and Gettysburg College for nurturing my love of teaching. I am forever indebted to Ambassadors Susan Burk and Norm Wulf for mentoring me in the ways of nuclear nonproliferation, and along with others in the former Arms Control and Disarmament Agency for giving me a truly memorable experience as we worked tirelessly in 1994 and 1995 to secure the indefinite extension of the Nuclear Nonproliferation Treaty.

I want to especially thank Publisher Joel Claypool of Morgan & Claypool Publishers for getting me started on this project, and the language and production editors, led by Jacky Mucklow, for their professional assistance in producing this book and significantly improving its quality.

Lastly and most importantly, I want to thank my dear wife Eleanor, whose love and tireless support makes everything else possible.

Author biography

Peter Pella

Dr Pella has just retired after being a Physics professor for over 35 years. He spent the last 28 years at Gettysburg College where he was the W K T Sahm Professor of Physics. His research included the study of the spin response of the nuclear force and fundamental properties of the neutron. He has participated in research at the Indiana University Cyclotron Facility, the Bates Linear Accelerator Facility and the Thomas Jefferson Continuous Electron Beam Accelerator facility in medium-energy nuclear physics. He is also involved in issues related to nuclear weapons. His expertise focuses on nuclear nonproliferation, the Nuclear Nonproliferation Treaty, The International Atomic Energy Agency, and nuclear issues involving North Korea and Iran.

As a William Foster Fellow from 1994 to 1995, he worked at the United States Arms Control and Disarmament Agency, from which he received a Meritorious Honor Award for his service in helping to achieve the indefinite extension of the Treaty on the Non-Proliferation of Nuclear Weapons. He also worked at the Bureau of Nonproliferation, US Department of State, from 2000 to 2001 on issues relating to Iraq, North Korea, and The International Atomic Energy Agency. He contributed two chapters (Nuclear Nonproliferation and The International Atomic Energy Agency) to the Oxford International Encyclopedia of Peace published by Oxford University Press in October 2009. He also authored a textbook, *Nuclear Weapons, Policy, and Strategy, The Uses of Atomic Energy in an Increasingly Complex World*, for a nuclear weapons policy course he had taught for over 30 years.

He holds a bachelor's degree in nuclear engineering from the United States Military Academy at West Point; a master's degree in experimental nuclear physics from Rensselaer Polytechnic Institute; and a doctorate in experimental nuclear physics from Kent State University.

IOP Concise Physics

The Midlife Crisis of the Nuclear Nonproliferation Treaty

Peter Pella

Chapter 1

Introduction

The nuclear age officially began seventy years ago in the desert outside of Alamogordo, New Mexico. A weapon test, called 'Trinity', demonstrated the feasibility of constructing nuclear weapons, with their awesome power. Two more explosions quickly followed that test. They were detonated over the cities of Hiroshima and Nagasaki and ended World War II. Thankfully those are the only times that such weapons have been used in anger, but the effects of those explosions left an indelible mark on the psyche of the world's population.

The sense of idealism felt by the victorious countries after World War II ended quickly eroded due to the constant friction between the United States and the then Soviet Union (also known as the Union of Soviet Socialist Republics or USSR). The term 'Cold War' was coined in 1947 to describe the continuing conflict between the countries. The world became divided between those states controlled by the Soviet Union (the Warsaw Pact) and those allied with the US (NATO).

The first detailed nuclear doctrine for the United States was issued in 1948. It stated that US national security and international stability depended on containing any Soviet expansion. To do so, the US needed to develop a retaliatory nuclear force in case it was attacked first by nuclear weapons, and to build more weapons for a pre-emptive strike. Because of the large advantage that the Warsaw Pact had over NATO in conventional forces, it also recommended that nuclear forces be used in any conventional war, if the US found itself at a disadvantage. This was an effective policy at the time because of the US monopoly on nuclear weapons and was called 'massive retaliation'.

During the late 1950s, the policy began to shift. In 1957 the US had some 4000 weapons and the USSR had about 500. The US realized that the USSR would soon have a sizeable bomber and intercontinental ballistic missile (ICBM) capability. The massive retaliation policy was no longer effective. If the US tried to destroy the USSR, the USSR would surely do the same to the US with its surviving nuclear weapons.

Only a few weapons detonated over major US cities would be disastrous. Therefore a different policy of deterrence was developed.

This new doctrine became known as mutual assured destruction (MAD). It was designed to deter a nuclear attack against the US and provide the same deterrence for the USSR. MAD would prevent a surprise nuclear attack against the US in the following way. The US would ensure that if they were attacked first (in a 'first strike') and some of their nuclear weapons capability were destroyed, there would be enough survivable nuclear forces left to be able to effectively destroy the USSR. Such a retaliatory capability (used for a 'second strike') would deter the USSR from striking first. Because of Soviet nuclear forces, the same deterrence would work to prevent the US from launching a first strike against the USSR. Deterrence would be mutual.

This doctrine required two components that were particularly worrisome to civilian populations around the globe. First of all, it required nuclear forces that were on hair-trigger alert, so that a second strike could be carried out in 30 min or less. It also required the US to publicly proclaim that it not only had the capability, but also the will to carry out such a second strike. The fear that a catastrophic nuclear war between the US and the USSR could be started by accident or by a rogue military commander was palpable.

Hollywood and popular literature addressed these fears. Two movies were particularly effective, though there were many others that followed suit. *Fail Safe* was about a lone nuclear-armed bomber that attacked Moscow because of a technical accident, and the black comedy *Dr Strangelove or: How I Learned to Stop Worrying and Love the Bomb,* starring Peter Sellers, was about a nuclear attack against the USSR launched by a rogue commander. In addition, atmospheric testing by the US and USSR as they worked to develop hydrogen weapons and perfect their fission bombs raised concerns about the effects of the fallout produced from such tests. A popular book at the time by Neville Shute, called *On the Beach*, later made into a movie of the same name, depicted the last days of humanity after a catastrophic nuclear war. The last survivors on Earth were in Australia, and they were waiting for the lethal radioactivity that was moving their way.

It became clear to the international community that nuclear weapons were becoming a significant threat to international security. Nuclear reactor technology was rapidly spreading around the globe, thanks in part to the Eisenhower administration's 'Atoms for Peace' program. Some countries were then using those technologies to develop nuclear weapons and forecasters estimated that within the next several years there could be as many as 20 or 30 nuclear weapons states. The Nuclear Nonproliferation Treaty (NPT) was formulated to respond to that threat.

The NPT entered into force 46 years ago with 45 original signatories, including the nuclear-armed states Russia (then known as the Soviet Union), the United Kingdom and the United States. This was preceded by three years of tense negotiations among 18 nations in Geneva and two more years of waiting while the world watched to see if other countries would sign and ratify the treaty. Finally, on 5 March 1970 the treaty became part of the international landscape.

The NPT represents one of the most important bargains ever negotiated. States without nuclear weapons pledged not to acquire them, while nuclear-armed states committed to eventually give them up. At the same time, the NPT allowed the peaceful use of nuclear technology by non-nuclear-weapon states under strict and verifiable control. It is the most widely subscribed arms control treaty and represents an international norm against nuclear weapons proliferation.

The existence of the NPT has led several states to abandon their nuclear weapons ambitions and it has made it far more difficult for other non-nuclear-weapon states to acquire the materials and technology needed to build them. The NPT process has also encouraged action on several nuclear arms control initiatives. It has helped promote regional security by giving assurances to countries within a certain region that their neighbors are not developing nuclear weapons. This helps to reduce the incentives for others to seek nuclear arms for prestige or defense.

Although the NPT has survived for over 45 years, and has dealt with such issues as the Cold War, the dissolution of the former Soviet Union, a clandestine nuclear weapons program in Iraq, and a nuclear-armed North Korea (Democratic People's Republic of Korea), there are even more stresses on it today. There need to be even stronger and more comprehensive efforts on nuclear nonproliferation. The NPT's future success depends on universal compliance with stricter procedures to prevent the spread of nuclear weapons, more effective regional security strategies, especially in the Middle East, and renewed progress toward fulfillment of the nuclear-weapon states' NPT disarmament obligations.

Chapter 2

Scientific background

To understand the concerns raised by the proliferation of nuclear weapons and the details of nuclear nonproliferation efforts, one must have a basic understanding of some of the scientific principles involved. The intention of this chapter is to introduce the most important scientific principles in a manner accessible to most readers. Those with a more scientific background will realize that much has been left out, though hopefully not enough to affect one's understanding of the rest of the material in this book.

2.1 Fundamental forces

There are four naturally occurring forces in the Universe: (1) the gravitational force, (2) the electric force, (3) the strong nuclear force, and (4) the weak nuclear force. They are called the four fundamental forces and are responsible for all that exists in nature and for life itself.

(1) The gravitational force is an attractive force between two objects that have mass. It is the force responsible for holding the Earth in its orbit around the Sun, allowing the Earth to receive the appropriate temperature for life to exist. Gravity is responsible for attracting the elements together, forming the Earth and Sun. It also governs the motions of all the Universe's stars, planets, and galaxies.

(2) The electric force is a force between two objects that have a property called charge. There are two kinds of charges and Benjamin Franklin labeled them positive and negative. If the two objects possess the same sign of charge (both positively or negatively charged), then the force is a repulsive one. If the two objects have charges of opposite sign, then the force is an attractive one. When an object contains an excess of positive (negative) charge, we say that it is positively (negatively) charged. If an object contains equal amounts of positive and negative charges, we say that the object has zero net charge or is neutrally charged. The electric force is responsible for holding atoms together. It is the force responsible for interactions between

molecules, and hence all chemical reactions. Since life itself is a process of chemical reactions, the electric force is responsible for life.

(3) There are two types of nuclear forces. The first type is the strong nuclear force. It is responsible for holding the nuclei of atoms together, for fission reactions, and for fusion reactions. Fission alone, or both fission and fusion together, are responsible for the energy released in nuclear weapons. Fission is also the process by which the energy is released in a nuclear reactor. Fusion is responsible for the Sun's heat generation, which provides the Earth with the proper temperature and light to sustain life. Finally, fusion in stars is also responsible for producing elements within the periodic table up to iron.

(4) The second type of nuclear force is called the weak nuclear force. This force is responsible for, among other things, a type of radioactive decay called beta decay. Beta decay is responsible for producing all the elements in the periodic table from iron up to uranium. These elements are crucial for the formation of planets, and are necessary to support life on Earth.

2.2 Model of the atom

An atom is the smallest piece of an element that contains all that element's properties. The electric force between the negatively charged electrons orbiting the atom and the positively charged protons in the nucleus holds it together. A neutral atom is one in which there are equal positive and negative charges.

The nucleus contains most of the atom's mass. The size of the nucleus is on the order of 10^{-15} m, while the size of the atom itself is on the order of 10^{-10} m. The nucleus, therefore is 1/100 000th the size of the whole atom. To put that in perspective, if the nucleus were magnified to be the size of a small marble, and we placed that marble at second base in the Royals Stadium[1], the outer electrons would be flying around in the bleachers and outside the stadium. An atom is mostly empty space.

The nucleus is made up of two kinds of particles: the proton and the neutron. The proton has a positive charge equal in magnitude to the negative charge on an electron. A neutral atom, therefore, contains equal numbers of protons and electrons. The neutron has zero net charge. Table 2.1 summarizes the characteristics of an atom's three types of particles. Note that the masses are listed in both

Table 2.1. Particle masses.

Particle	Mass (kg)	Mass (μ)
Electron	9.11×10^{-31}	5.485×10^{-4}
Proton	1.672623×10^{-27}	1.0072765
Neutron	1.674928×10^{-27}	1.0086649

---

[1] A baseball park in Kansas City, Missouri, USA.

kilograms and atomic mass units. Because these particles' masses are so small, a unit of mass was developed by chemists and called an atomic mass unit (μ). An atomic mass unit is defined as 1/12th of the mass of a carbon atom, so that $1\mu = 1.66054 \times 10^{-27}$ kg. The mass of the proton is almost equal to that of the neutron and the mass of the electron is 2000 times smaller than either one.

The periodic table of elements lists all the known elements in order of the number of protons in the nucleus (which is also the number of electrons in the neutral atom). That number is called the atomic number and is represented by the letter Z. Elements in the same column of the periodic table have similar chemical properties because they have similar outer electron configurations. The periodic table's elements, listed by their chemical symbols and Z values, are shown in figure 2.1.

In order to describe a particular nucleus one needs to develop a shorthand way of referring to them. A nucleus is denoted by its atom's chemical symbol and three numbers—the atomic number (Z), the number of neutrons in the nucleus (N), and the atomic mass number (A). A is also the number of particles in the nucleus, so that $A = N + Z$. If 'El' represents an element, we designate a nucleus in the following manner:

$$^{A}_{Z}\text{El}_{N} \rightarrow {}^{A}_{Z}\text{El} \rightarrow {}^{A}\text{El}.$$

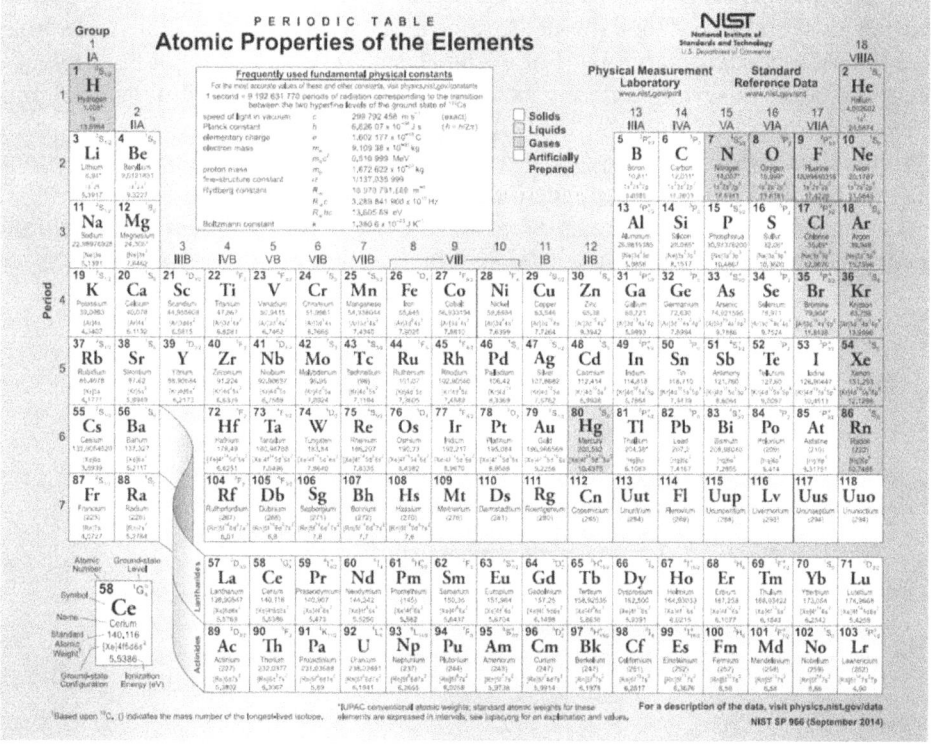

Figure 2.1. The periodic table of the elements. Reproduced courtesy of the Nuclear Regulatory Commission.

You should note that since $A = N + Z$ and the Z is determined by the element's location in the periodic table, the first two expressions contain redundant information. An example would be an isotope of carbon. Since the location of carbon in the periodic table tells us that $Z = 6$ and this isotope of interest has seven neutrons, we have:

$$^{13}_{6}C_7 \rightarrow {}^{13}_{6}C \rightarrow {}^{13}C$$

and we refer to this nucleus as 'carbon-13'.

Since the atom's mass is primarily due to the nucleus, and the mass of each proton and neutron is approximately equal to one atomic mass unit, an atom's mass in atomic mass units can be approximated by the atomic mass number A. As was stated before, an element's chemical properties are determined by the atom's number of electrons, and in particular the outer electrons. The nuclear properties of a nucleus are determined by the number of neutrons and protons within the nucleus and their configuration.

2.3 Isotopes

The nucleus of a particular element can exist in a number of different orientations. These nuclei all have the same proton number Z but a different neutron number N. They are called isotopes. Isotopes of the same element have differing masses, but the atoms that contain them all have the same chemical properties. This is because they all have the same Z, and hence the same number and configuration of electrons.

Some isotopes exist naturally. For example, lead (chemical symbol Pb) present in the soil consists of several different isotopes. Table 2.2 shows the different naturally occurring isotopes of lead and their abundances (as percentages). The lead in the apron that one wears when having an x-ray taken is composed of these four isotopes of lead, but they are chemically identical.

Two important isotopes for this discussion are two of the main isotopes in naturally occurring uranium (U). They are listed in table 2.3, along with their abundances and

Table 2.2. Isotope abundances.

Isotope	Abundance
$^{208}_{82}Pb_{126}$	52.3%
^{207}Pb	22.6%
^{206}Pb	23.6%
^{204}Pb	1.5%

Table 2.3. Uranium isotopes.

Isotope	Abundance	Mass (μ)
$^{238}_{92}U_{146}$	99.27%	238.0507826
^{235}U	0.72%	235.0439231

masses in atomic mass units. They are both chemically identical but have different masses. Also note that their masses in atomic mass units are almost equal to the atomic mass number A, also in atomic mass units.

2.4 Stability

Some naturally occurring isotopes, such as the uranium isotopes in table 2.3, and some artificially produced isotopes are radioactive and change into other nuclei. Other isotopes do not change, such as the lead isotopes in table 2.2, and are called stable isotopes.

A gentleman named Albert Einstein said that mass and energy are equivalent. This means that we can convert mass into energy (as in a reactor) and energy into mass (in an accelerator). If an isotope's mass is lower than any other possible configuration of the neutrons and protons within that isotope, then that isotope is stable. If there is another configuration of neutrons and protons with less mass than the original isotope, that isotope will rearrange itself to that lower mass configuration. This is called radioactive decay. As an isotope decays from a more massive state to some combination of neutrons and protons with less mass, that small difference in mass is converted into energy.

2.5 Half-life

Naturally occurring radioactive substances will decay by emitting radioactive particles (alpha, beta decay, or gamma) and change into a different isotope that may or may not be radioactive itself. However, if a radioactive nucleus were sitting in front of someone, that person would not be able to determine the exact instant that the nucleus would decay. All one can determine is the probability that it will decay in the next second. This is due to the fundamentally probabilistic nature of the theory scientists use to describe nature on the atomic and nuclear level: quantum mechanics. Every radioactive nucleus has a unique probability for decaying.

One way to describe this probability is to use the half-life ($T_{1/2}$). The half-life is defined as the time it takes for a particular amount of a radioactive substance to decay down to about ½ of its initial amount. After two half-lives there will be ¼ of the original amount, and so on. Since each radioactive isotope has a unique probability for decaying, it also has a unique half-life. For example, the half-life of $^{238}_{92}U_{148}$ is 4.5×10^9 years and the half-life of $^{14}_{6}C_{8}$ is 5730 years.

2.6 Fission and fusion

Fission is a nuclear process that occurs when a nucleus with an atomic mass number (A) that is very large (usually <200, called a heavy nucleus) breaks apart into two smaller or lighter nuclei, with an A of about 100 for each of them. Fission is usually accompanied by the emission of neutrons and the two lighter nuclei are radioactive.

There are two kinds of fission processes. Spontaneous fission occurs when a heavy nucleus breaks apart on its own, as opposed to other forms of radioactive decay. The second form of fission is called induced fission. In this process, the fission of a heavy nucleus is preceded by the absorption of a neutron.

An example of an induced fission reaction is:

$$_0^1n_1 + {}^{235}_{92}U \rightarrow {}^{141}_{56}Ba + {}^{92}_{36}kr + 3(_0^1n_1).$$

When the ${}^{235}_{92}U$ nucleus absorbs an incoming neutron, represented by the symbol $_0^1n_1$, it begins to oscillate. It eventually becomes unstable and splits apart into two lower mass nuclei, ${}^{141}_{92}Ba$ and ${}^{92}_{36}Kr$, called fission products, releasing three neutrons and energy. The energy released represents about 20% of the mass of one proton or neutron and can be calculated by taking the difference between the masses on the left and right sides of the reaction and converting these to energy using Einstein's famous formula $E = mc^2$, where m is the difference in masses, E is the energy released and c is the speed of light. Since c is a very large number and c^2 is even larger, a small mass difference can result in a large release of energy.

The ${}^{235}_{92}U$ nucleus does not produce the same number of neutrons or the same fission products every time it undergoes a fission reaction. Just like radioactive decay, the fission reaction is governed by the probabilistic nature of quantum mechanics. Figure 2.2 shows the percentage of fission products produced in a fission reaction involving ${}^{235}_{92}U$ and their atomic mass numbers for different fissionable isotopes. The probabilities peak at around $A = 95$ and $A = 137$, so that the two fission products produced are not of equal mass.

The fission products are highly radioactive and will undergo several decays (emitting radioactive particles) until they eventually become stable isotopes. When a large amount of ${}^{235}_{92}U$, undergoes fission, the amount and variety of radioactive fission products is also large, and the radiation emitted remains significant for a long time.

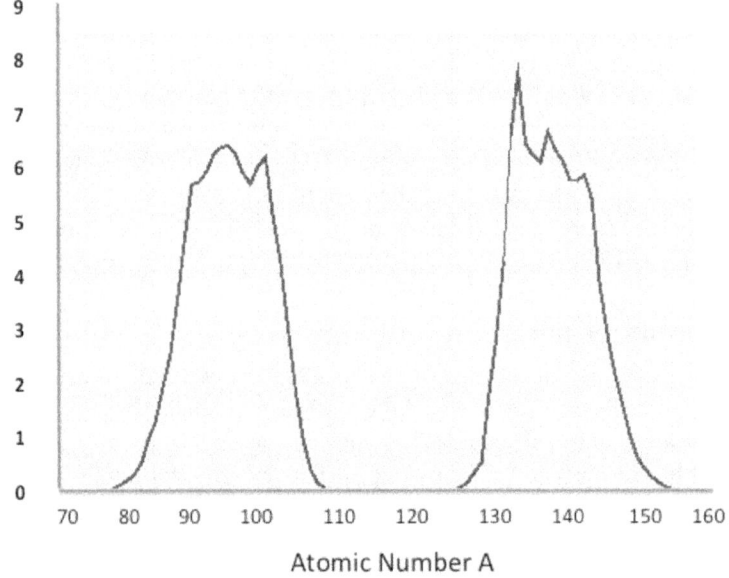

Figure 2.2. Fission fragments. Reproduced courtesy of Lawrence Livermore Laboratory.

The energy released in a fission reaction is much larger than in a typical chemical reaction. The ratio is about five million. This means that for equal amounts of fuel, a fission reaction releases about five million times more energy. As an example, one kiloton (1 kT) of trinitrotoluene (TNT), which is 1000 metric tons of TNT (1 ton = 1000 kg), when detonated releases the same amount of energy (via a chemical explosion) as is released when 0.056 kg of uranium undergoes fission. In view of this, you can appreciate the commercial and military interest in studying fission.

Since each fission releases neutrons, the neutrons from a previous fission reaction can cause a subsequent fission reaction, resulting in what is called a chain reaction. As an example, let us assume that two neutrons from a previous fission reaction are available to cause future fission reactions. The first fission reaction is then followed by two fission reactions and those two fission reactions are followed by four fission reactions, and so on.

After 80 iterations of this process (called generations) there will be 2^{80} fission reactions, or 1×10^{24} fission reactions. This whole process takes place in about 1 µs, resulting in an explosion that blows the remaining material apart and releases the energy equivalent of an explosion of 10 kT of TNT. The last ten generations release almost 99.9% of the energy and this happens within 10^{-8} s. This is what happens in a nuclear weapon. In a nuclear reactor that relies on fission to produce energy, one would like to ensure that only one neutron from a previous fission is used to cause the next fission.

IOP Concise Physics

The Midlife Crisis of the Nuclear Nonproliferation Treaty

Peter Pella

Chapter 3

Technology

3.1 Reactor physics

Except when solar energy is used, one generates electricity by rotating large coils of wire that sit inside a strong magnetic field in a device called a turbine. This process produces the alternating current that is used in households and industry around the world. The coils are made to rotate in several different ways. They can be rotated by falling water, as in a hydroelectric dam, or wind can turn propellers connected to the coils, as in a wind generator. The most common way to rotate the coils is to force steam, at very high pressure, through the turbine. The coils are connected to blades and the steam causes the blades and coils to turn. This steam is derived from boiling water and the water can be heated in a number of ways. The most common way to generate the heat is to burn coal, oil, or natural gas.

A nuclear reactor uses induced fission to generate heat. A nuclear reactor then uses the heat to produce steam. Beyond this point, the nuclear reactor is no different from any other power plant that burns coal, oil, or natural gas to produce heat and create steam to generate electricity as stated above. Although the designs and materials for nuclear reactors can differ greatly, they all operate in a similar manner.

There are four major components of a commercial nuclear reactor: the fuel, moderator, coolant, and control rods. Reactor design is quite complex. The fission reactions in a nuclear reactor must be designed so that the energy is produced continuously in a controllable manner. A way to represent that is with a number called the k constant. If $k = 1$, then only one neutron from the previous fission will cause a subsequent fission. If $k < 1$, then less than one neutron is available to cause a subsequent fission and the reaction will die out. If $k > 1$, then more than one neutron is available to create a subsequent fission and the reaction will produce too much energy and will be difficult to control.

The k factor depends on a number of design elements. The shape of the fuel affects how many neutrons escape the fuel without causing a fission reaction. The types of materials used in the reactor determine how many neutrons are absorbed by these materials and are unavailable for fission. Finally, the type and composition of the fuel determine how many neutrons are produced per fission and how many are absorbed without causing a fission reaction. A reasonable start would be to use natural uranium with a small percentage of $^{235}_{92}U$ as the fuel. The neutrons emitted in a fission reaction are moving very fast and quite understandably are called fast neutrons. However, in a nuclear reactor these fast neutrons are slowed down before they cause a subsequent fission. Since $^{238}_{92}U$ does not undergo fission when it absorbs a slow neutron, it is the fission reaction in the small amount of $^{235}_{92}U$ in natural uranium that produces the desired energy release.

The moderator in a nuclear reactor functions to slow down the neutrons. The fast neutrons lose energy, and hence slow down, by colliding with the nuclei of the moderator. A fast moving object loses the most energy in a collision when it collides with a stationary object of nearly the same mass. This makes sense if one thinks about the head-on collision between two billiard balls. The incoming ball stops after the single collision. Now think about what happens when an air gun pellet is shot at a much larger mass bowling ball. The BB hardly slows at all, ricocheting off at almost the same speed.

A good moderator should also not absorb the neutrons that are needed to keep the fission reactions going. The smallest mass atom is, of course, hydrogen; and hydrogen, as part of the water molecule, is used as a moderator. However, hydrogen has a small probability of absorbing a neutron, which removes it from the chain reaction process. Therefore, when water is used as a moderator, the nuclear reactor fuel cannot be natural uranium (which contains 0.7% of $^{235}_{92}U$). The reactor fuel must be 'enriched' to contain 3–5% of $^{235}_{92}U$.

Deuterium is an isotope of hydrogen which has a neutron in the nucleus along with a proton and its symbol is $^{2}_{1}H$. Deuterium can be used, again in the form of water, called heavy water, as a moderator. Deuterium does not absorb neutrons as easily as hydrogen, so a reactor with heavy water as a moderator can use natural uranium as its fuel. Carbon, which is heavier than hydrogen or deuterium, can also be used as a moderator. A reactor with carbon, in the form of graphite, as a moderator also uses enriched uranium as a fuel.

A critical part of a reactor is the coolant. The coolant removes the heat from the reactor core, which contains the nuclear fuel, and transfers the heat in some manner to create steam. If one uses water or heavy water as the moderator, that moderator can also serve as the coolant. The heat in the reactor core is not only generated by the fission reactions, but also by the heat generated from the radioactive decay of the fission byproducts. Therefore, even if the fission reactions have stopped, coolant is needed to remove the heat from the fission byproducts. The loss of coolant is one of the most serious concerns associated with nuclear reactor safety. Without the coolant, the reactor core can become hot enough to melt the fuel, melt through the reactor building, and melt into the ground. (This is what some call the 'China Syndrome'; see also the movie of the same name.)

One also needs something else to absorb any extra neutrons present so that $k = 1$. Control rods help to control the number of neutrons available for fission and are generally inserted part way into a reactor. They are connected to motors that can move them further into or out of the reactor. If too many neutrons are present, the reactor produces too much heat and is in danger of overheating, which would melt the reactor fuel. The control rods are inserted further into the reactor, absorbing more neutrons and reducing the power production. If not enough neutrons are present, then the control rods are pulled part way out of the reactor so fewer neutrons are absorbed. This ensures that $k = 1$ at all times. A good element to use for control rods is one that readily absorbs slow neutrons. The two most widely used elements are cadmium and indium.

Fission reactions happen very quickly. The mechanical motion of the rods is too slow to regulate the number of neutrons that are produced directly from fission. Even a k of 1.01 is high enough for the chain reaction to get out of hand. Luckily there are 'delayed neutrons'. Some of the products produced in fission decay by emitting neutrons. They have half-lives on the order of 1–10 s. These delayed neutrons are the key to controlling the activity within a reactor. A reactor is designed so that $k \approx 0.95$. The other 0.5 comes from the population of delayed neutrons. The control rods then respond to these delayed neutrons, maintaining $k = 1$ for the reactor.

There is an important safety feature associated with the control rod system. This system is usually placed above the reactor core. The motors that move the control rods hold them in place through the use of electromagnets. Should the main power to the reactor be cut off for some reason, the electric current flowing through the electromagnets stops and the control rods are no longer held in place. Gravity causes the control rods to fall completely into the reactor core, absorbing lots of neutrons and effectively shutting down the fission reactions in the reactor. However, the reactor core still needs to be cooled because of the heat generated by the fission byproducts.

3.2 Nuclear reactors

Figures 3.1, 3.2 and 3.3 show the number of nuclear reactors currently in operation in the world (by country), how much of each country's electrical power is supplied by nuclear reactors, and the number of reactors under construction.

Most nuclear reactors generate about 1000 MW of electrical energy. This is enough energy for a city of about one million people (not including industrial use). Table 3.1 shows the amount of fuel needed to generate 1000 MW of electrical power for coal-fired and nuclear power plants. It should be noted that producing electricity from steam is only about 33% efficient. This means that a power plant that produces 1000 MW of electrical power (written as 1000 MW(e)) must produce 3000 MW of thermal energy (written as 3000 MW(th)).

One can see the attractiveness of using uranium as fuel. Much less fuel is required than when trying to burn fossil fuels. There is also no emission of greenhouse gases or other pollutants from a nuclear reactor. Of course the major downsides to nuclear reactors are the public fear of accidents and anything radioactive, and the problem of what to do with the nuclear waste, which has still not been solved after more than 50 years.

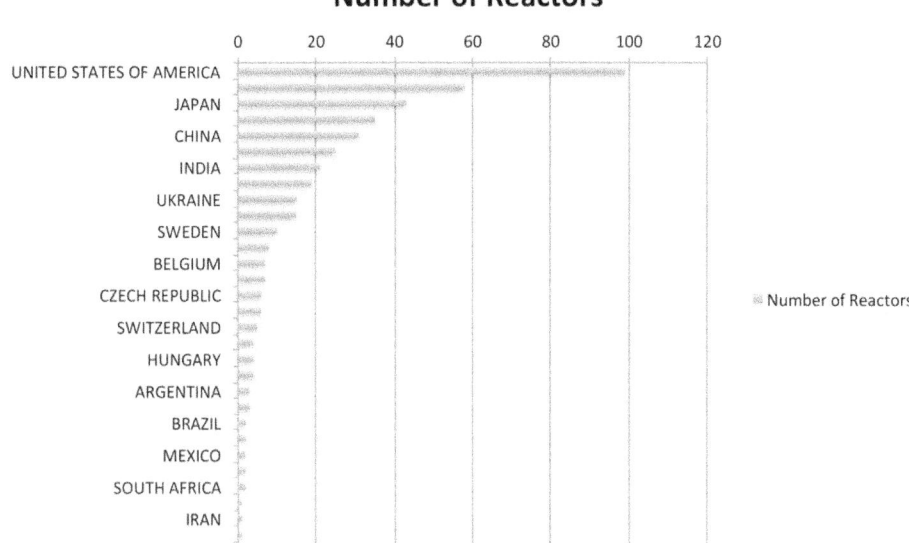

Figure 3.1. Worldwide nuclear reactors. Reproduced from [1].

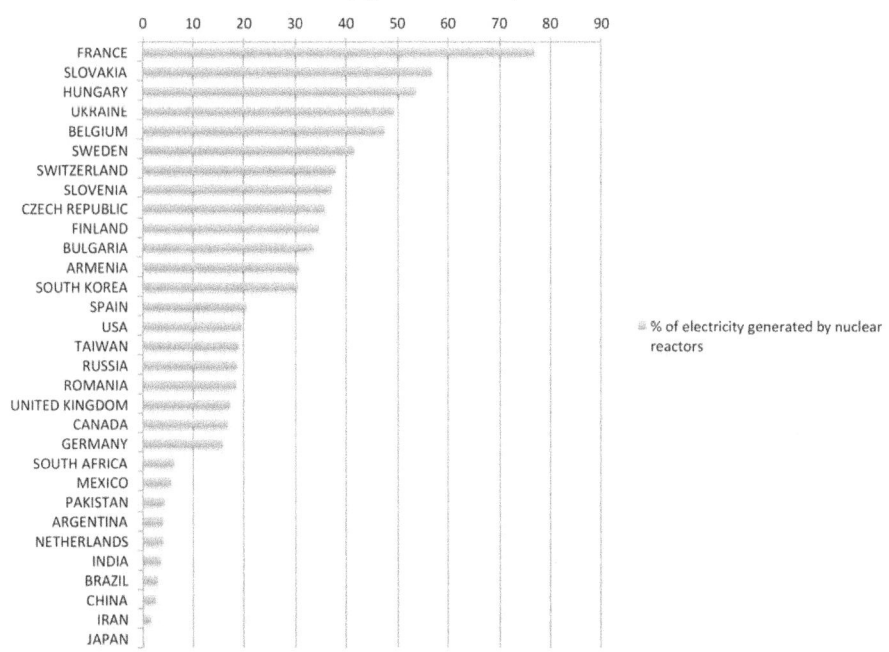

Figure 3.2. Percentage of energy supplied by nuclear reactors. Reproduced from [2].

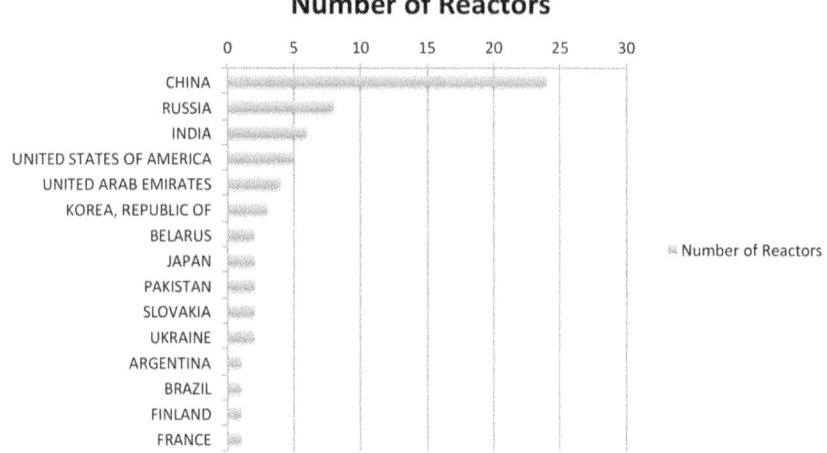

Figure 3.3. Number of reactors under construction. Reproduced from [3].

Table 3.1. Data taken from [4].

Fuel used during operation	Fuel needed to run 1000 MW(e) coal-fired and nuclear power plants		
	Time of operation		
	1 h	1 day	1 year
Coal plant	350 t	8.4×10^3 t	3.0×10^6 t
Nuclear plant	0.3 lb	7.2 lb	1.3 t

Besides those in Russia, China, and Japan, most new reactors are being built in developing countries. China, India, and Ukraine account for almost half of the reactors under construction in the world today. Many of the less robust economies are affected greatly by fluctuations in oil prices. In addition, many developing countries receive financial and technical assistance from the US, the International Atomic Energy Agency, and other countries for their reactor programs. Some countries find that having an indigenous nuclear power industry provides them with a technology infrastructure that helps with other modernization needs. Concern about climate change from greenhouse gas emissions is also a driving force.

3.3 Reactor designs

Most reactors in the world use ordinary (light) water as both the coolant and the moderator. The water surrounds the reactor core in the vessel and fills the spaces between the fuel rods in the fuel assembly. Since water absorbs some of the neutrons produced in the fission process, the reactor fuel is enriched to contain 3–5% of $^{235}_{92}U$.

There are two kinds of light water reactors. In a boiling water reactor (BWR), the water in the reactor vessel removes the heat from the core and begins to boil.

The steam produced by the boiling water is collected and flows through the turbines that generate the electricity. The steam then flows through a condenser. The condenser uses a source of outside water, usually a river or the ocean. The outside water absorbs the heat remaining in the steam, changing it back into water, which then flows into the reactor vessel to act as the moderator and coolant once again in a continuous loop.

The outside water is returned to the source. If the source of the outside water comes from a river or the ocean, great care is taken to ensure that the water is returned to the source at the same temperature at which it was extracted, minimizing the environmental impact. Before being returned to the source, the water is pumped up to the top of the large asymmetrical hourglass-shaped towers that are the most prominent features of nuclear power plants. This shape is called a venturi. Venturi shapes are also used in the carburetors of cars and for airfoils. As the water trickles down the sides of the towers, it cools by evaporation, giving off the water vapor that is usually seen rising from the towers. About one-third of the coolant water is released into the atmosphere by this cooling process.

One concern about BWRs is that the steam produced inside the reactor vessel is slightly radioactive, due to the water being in contact with the reactor core. It is difficult to design a steam turbine that is impervious to small leaks of the steam. The pressurized water reactor (PWR) addresses this problem by adding a second coolant loop. The primary loop, which is inside the reactor vessel, consists of pressurized water. The water again acts as the coolant and moderator, but it is under high pressure. This allows the water in the primary loop to obtain high temperatures without boiling. A secondary loop, consisting of water under atmospheric pressure, removes the heat from the primary loop, causing the water in the secondary loop to boil, producing the steam. The heat is removed from the water in the secondary loop using an outside source of water, as described above.

Canada has developed a reactor design that uses heavy water, under pressure, as the coolant and moderator. Since deuterium has a low probability of absorbing neutrons, one can use natural uranium as the fuel. It is called the Canadian deuterium uranium (CANDU) reactor. Canada has a number of large hydroelectric power plants that can be used to create heavy water. They therefore do not have to develop an enrichment capability.

Another type of reactor is the graphite reactor. A graphite reactor uses carbon, in the form of graphite, as the moderator. Most of the graphite reactors in the world use light water as the coolant and are similar in operation to the boiling water reactor. The use of water as the coolant means that the fuel must be enriched uranium. Other graphite reactors operate at much higher temperatures and are cooled by gas (helium or CO_2).

There has been much research in the US and other countries (particularly Russia) on advanced nuclear reactor designs for both light water and graphite reactors. The focus is on added safety features and simpler designs. The Russians currently have 50 of their advanced light water reactors in operation. There are another two advanced light water reactors and 14 advanced graphite reactors in operation in other parts of the world. Two excellent web-based sources for all the issues described are the IAEA[1] and NRC[2] websites.

There are three breeder reactors in operation in the world today. A breeder reactor is a nuclear reactor that uses sodium as a coolant. Sodium liquefies at 98° C and does not boil until it reaches 892° C, allowing the reactor to operate at very high temperatures. Sodium is very corrosive, so great care must be taken to ensure the coolant system remains intact. A breeder reactor uses uranium enriched to 15–30% $^{235}_{92}U$, which allows the fission reaction to occur using fast neutrons, without needing a moderator.

A blanket of $^{238}_{92}U$ surrounds the reactor core. By the time the neutrons from inside the reactor leave the core, they are moving slowly. The $^{238}_{92}U$ then absorbs these neutrons. As the $^{238}_{92}U$ nucleus absorbs one neutron, it changes into $^{239}_{92}U$. This isotope is radioactive and eventually decays to $^{239}_{94}Pu$, which has a half-life of about 24 000 years. The blanket is then removed from the reactor and the plutonium is extracted through chemical reprocessing. The reactor is designed to then either use the extracted plutonium in its core as fuel or continue with the same enriched uranium fuel to produce more plutonium. Current breeder reactors can produce more plutonium in the blanket than fuel is burned in the reactor core (by about a factor of 1.2, although they are designed to produced 1.4 times the amount of fuel used). One can then rely on nuclear reactors as a source of energy production long after all the $^{235}_{92}U$ existing in the world has been used, because there is about 140 times more $^{238}_{92}U$ than $^{235}_{92}U$.

France and Japan have been operating breeder reactor programs. They are planning to redesign their other nuclear reactors to be able to burn a mixture of plutonium and uranium fuel (known as mixed oxides (MOX) fuel). However, the Japanese are reconsidering their options because of the accident at Fukushima. The US, under a directive from the Carter administration, stopped its research on breeder technology in the 1970s. The decision was based partially on economic concerns and partly on the fear of nuclear proliferation. The US has a large coal reserve, and President Carter felt that uranium resources would be adequate for a long time to come. A breeder reactor program means that there will be stockpiles of plutonium controlled by the commercial reactor companies and President Carter and Congress felt that plutonium would be less secure when not under military control, and that it would be more likely to be stolen by terrorist groups or other countries wishing to build nuclear weapons.

However, when the Cold War ended and the US and Russia started dismantling significant parts of their nuclear arsenals, there was the problem of what to do with the plutonium taken from nuclear weapons. Since no easy answer has been found, the Clinton and George W Bush administrations reconsidered the option. The US Department of Energy is currently looking at ways to alter current nuclear power reactors so that they are able to burn MOX fuels.

---

[1] www.iaea.org.
[2] www.nrc.gov.

3.4 The nuclear fuel cycle and nuclear proliferation

The nuclear fuel cycle is the description of how uranium is made available for nuclear reactors and how the waste from burning it is handled. For this discussion, uranium enrichment technologies and nuclear spent fuel technologies are the most worrisome from the perspective of nuclear proliferation. In promoting the construction of nuclear reactors worldwide, one must confront the intimate relationship between nuclear reactor infrastructure and a nuclear weapons program.

3.5 Uranium enrichment

Natural uranium contains only 0.7% $^{235}_{92}U$. In most nuclear reactors, especially those in the United States, the uranium must be enriched to 3–5% $^{235}_{92}U$. As was mentioned earlier, since all uranium isotopes have the same chemical properties, one cannot use chemical techniques to separate $^{235}_{92}U$ from $^{238}_{92}U$. Most enrichment technologies exploit the mass difference between the 'lighter' (less massive) $^{235}_{92}U$ atoms and the 'heavier' (more massive) $^{238}_{92}U$ atoms.

Gaseous uranium hexafluoride (UF_6) is used in a number of the enrichment techniques. It is convenient to use because at atmospheric pressure it changes from a solid to a gas at 57° C, which is a relatively low temperature and easy to maintain. It does not exist as a liquid at atmospheric pressure, but it can be liquefied under pressure. Remember that when natural uranium is used to make UF_6, 0.7% of the molecules contain $^{235}_{92}U$.

Most enrichment facilities in the world use a device called a gas centrifuge. Uranium hexafluoride gas is spun around at extremely high speeds (20,000 rpm or higher). The heavier molecules move to the outside of the device and the lighter molecules are more numerous in the center. The lighter molecules are then extracted and sent into another centrifuge in a cascade like that shown in figure 3.4.

Figure 3.4. Reproduced courtesy of the US Department of Energy.

Depending on the size and sophistication of the centrifuges, it takes about ten stages of enrichment to go from natural uranium to 3%–5% enrichment. A total of 100 stages would be needed to produce an enrichment large enough to use in a nuclear weapon. Since each centrifuge only handles a small amount of material, several thousand centrifuges would be needed to produce enough enriched uranium to supply fuel for a large commercial reactor.

The technology required is very advanced. The centrifuges have to be machined to extremely high tolerances in order to spin so rapidly without tearing themselves apart. In addition, the UF_6 must be free of all contaminants. Finally, once the UF_6 is enriched to 3%–5% $^{235}_{92}U$, the uranium has to be converted to a solid to form the reactor fuel.

3.6 Reactor wastes

About one-third of the complete fuel for light water reactors is removed each year. The fuel that is removed is called 'spent fuel', but it still has usable fissionable isotopes in it. About a quarter of the original $^{235}_{92}U$ remains in the spent fuel in addition to the fission fragments, unused $^{238}_{92}U$ and isotopes of plutonium. Table 3.2 shows the percentage by weight of the various components of spent reactor fuel for a light water reactor [5].

A 1000 MW(e) nuclear reactor generates about 30 tons of highly radioactive wastes per year, which makes up a volume of about 60 ft^2. In the US, these wastes are generally stored at the bottom of deep, water-filled, steel-lined concrete pools on the reactor site. Water is circulated to keep the spent fuel from boiling the water because of the heat generated by the radioactive decay of the isotopes. However, most of the current nuclear power reactors have filled the original pools and are storing some of the wastes in above-ground concrete bunkers. There have also been larger pools constructed to house the wastes from several reactors in what are called 'away-from-reactor' pools. All these sites were designed as temporary waste storage pending the construction of a national waste storage facility, which has still not been constructed. A site was chosen for such a long-term storage facility and construction began in 2002 deep below Yucca Mountain in Nevada. However, based on concerns about the long-term stability of the site, the Obama administration cancelled the program.

Table 3.2. Approximate composition of spent fuel for a light water reactor.

Components	% by weight
$^{238}_{92}U$	95
$^{236}_{92}U$	0.4
$^{235}_{92}U$	0.8
Plutonium isotopes	0.9
Fission products	2.9

The need for long-term storage is important. Even after 100 years, the spent fuel from one reactor generates heat equivalent to over 250 100 W light bulbs. After 10 000 years, the radiation levels will be lethal to someone exposed to the spent fuel for a couple of hours. Obviously the current situation is inadequate.

3.7 Reprocessing

The recovery of the remaining uranium and the plutonium in the spent fuel is called reprocessing. The spent fuel rods are allowed to cool in pools of water for about five months. This also significantly reduces the radioactivity of the spent fuel due to the decay of a significant portion of the short-lived fission fragments. The fuel is then shipped to a reprocessing facility in specially designed and well-guarded shipping containers. The spent fuel rods are then cut up and dissolved in an aqueous nitric acid solution.

To understand fully the reprocessing process, one must have a detailed understanding of the chemical reactions and the valence states of the different nitrates, which is beyond the scope of this text. In short, through the use of organic solvents, the plutonium and uranium can be separated from the other fission fragments, which remain in the aqueous solution. The liquid waste, containing the still highly radioactive fission fragments, is allowed to dry, and then processed to form a glassy (borosilicate) substance that seems stable enough for long-term disposal.

The remaining uranium and plutonium material is further treated with nitric acid and organic solvents to separate the uranium from the plutonium. The standard process, called PUREX, results in an almost complete separation of the plutonium from the uranium isotopes. The uranium can then be sent back to an enrichment facility to produce more conventional reactor fuel, and the plutonium can be sent to a facility to produce MOX fuel. The UK, France, Japan, and Germany all have mature reprocessing technologies. Some countries, such as Switzerland and Belgium, send their spent fuel to other countries (in their cases, France) for reprocessing.

3.8 Fission weapons

To understand proliferation concerns, one must have a cursory knowledge of how a nuclear weapon works. One needs an uncontrolled chain reaction in a large amount of fissionable material for a fission weapon (also called an 'atom bomb'). That is, there must be more than one neutron from a previous fission to cause subsequent fission reactions that result in a dramatic release of energy. There are many ways that neutrons from a fission reaction can be lost so that they do not cause a subsequent fission in the fissionable material or fuel. They can escape from the fuel, or other, non-fissionable, material surrounding the fuel can absorb them.

The 'critical mass' is used to define how much fissionable material is needed for a nuclear weapon. The term critical mass is a bit of a misnomer because a critical mass does not only depend on the amount of fuel that we have, but also its shape, density, purity, and the material surrounding it. When we have less than a critical mass (subcritical mass), a sustained chain reaction cannot occur. In a weapon, there must

Table 3.3. Data taken from [6].

Fuel	Critical mass needed for a weapon		
	w/o tamper	Uranium tamper	Beryllium tamper
Natural uranium	—	—	—
20% $^{235}_{92}$U	160 kg	—	65 kg
50% $^{235}_{92}$U	68 kg	—	25 kg
100% $^{235}_{92}$U	47 kg	16 kg	14 kg
80% $^{239}_{94}$Pu	—	5.4 kg	—
100% $^{239}_{94}$Pu	10 kg	4.5 kg	4 kg

be more than a critical mass present (supercritical mass) for an explosion to occur. Table 3.3 shows the amount of fuel needed for a critical mass for various purities of $^{235}_{92}$U and $^{239}_{94}$Pu, the two nuclear materials of choice for nuclear weapons. The table also lists the effects of a tamper, which reduces the material needed for a critical mass. The main function of a tamper is to surround the fuel and reflect neutrons that might have escaped back into it to be used for subsequent fission reactions. Tampers are usually made from uranium or beryllium. Beryllium has the added feature of supplying extra neutrons via the following reaction:

$$^{1}_{0}n + ^{9}_{4}Be \rightarrow 2(^{1}_{0}n) + ^{8}_{4}Be.$$

Not much material is needed to produce a small nuclear weapon. A nuclear device made with 15 kg of uranium would be about the size of a grapefruit. If one used about 5 kg of plutonium, the weapon would be about the size of an orange.

3.9 Gun-barrel device

The nuclear weapon that was exploded over Hiroshima on 6 August 1945 used 50 kg of uranium enriched to 70% $^{235}_{92}$U. The spherical critical mass had a diameter of 6.7″. The overall weapon was 28″ in diameter, 10′ long and weighed 9000 lb. Its nickname was 'Little Boy' and it had a yield of 12.5 kT (equivalent to 12.5 kT of conventional explosive). The design for the weapon is called the 'gun barrel' and a sketch of the design is shown in figure 3.5.

Two separate pieces of the uranium fuel, neither one a critical mass, but both of them together more than a critical mass, are connected by a tube or gun barrel. Surrounding both fuel elements are tampers and some of the fuel is placed in front of a high-explosive charge. At the end of the gun barrel is a neutron trigger. Its purpose is to provide a burst of neutrons to start the fission process at the appropriate time. The high explosive detonates, shooting the smaller piece of fuel toward the rest of the bomb at very high speed. The two pieces of fuel unite, forming a supercritical mass. The neutron trigger releases neutrons to start the fission process, and an explosion occurs about 1 μs later. Currently, neutron triggers are made of small particle accelerators that accelerate tritium or deuterium nuclei and slam them into a

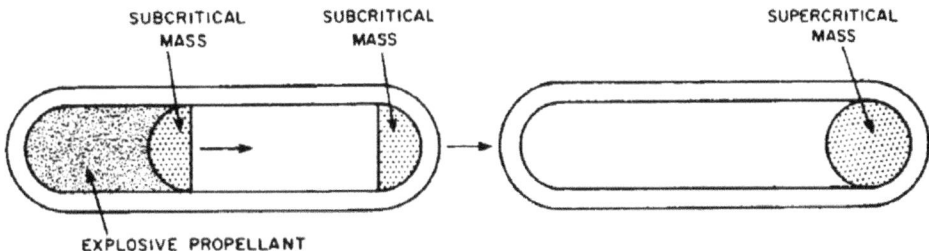

Figure 3.5. Reproduced with permission from [7] (p 15).

target made with deuterium. The tritium–deuterium reaction produces neutrons as shown below:

$${}^{2}_{1}H + {}^{2}_{1}H \rightarrow {}^{3}_{2}He + {}^{1}_{0}n.$$

The designers were so confident that this type of weapon would work that they did not test it before using it over Hiroshima.

3.10 Implosion device

The type of nuclear weapon that was first tested on 16 July 1945 in the desert outside of Alamogordo, New Mexico, and later used over Nagasaki on 9 August 1945, was an implosion device that used plutonium as the fuel. The weapon used over Nagasaki consisted of 5 kg of plutonium. It was 60″ in diameter, 10′ 8″ long, and was nicknamed 'Fat Man'. It had a yield of 20 kT.

As stated before, ${}^{239}_{94}Pu$ is produced in a reactor when natural uranium absorbs a slow neutron. However, ${}^{239}_{94}Pu$ can also absorb a neutron and change into ${}^{240}_{94}Pu$, and ${}^{240}_{94}Pu$ can absorb a neutron to become ${}^{241}_{94}Pu$. Therefore the plutonium extracted from a nuclear reactor will contain small amounts of ${}^{240}_{94}Pu$ and ${}^{241}_{94}Pu$ in addition to the ${}^{239}_{94}Pu$ needed for a nuclear weapon. Both ${}^{240}_{94}Pu$ and ${}^{241}_{94}Pu$ undergo spontaneous fission, which produces neutrons. The Manhattan Project scientists feared that the gun-barrel design would not be suitable for plutonium. It creates the supercritical mass needed for the explosion relatively slowly. The neutrons from the spontaneous fissions of ${}^{240}_{94}Pu$ and ${}^{241}_{94}Pu$ would start the fission of ${}^{239}_{94}Pu$ too early, before a supercritical mass was formed, and this would not achieve the large explosion desired, causing the weapon to 'fizzle' instead. A simple design for an implosion weapon is shown in figure 3.6.

A tamper and a specially designed array of high-explosive material surround a spherical, subcritical mass of plutonium. At the center of the sphere is a neutron trigger. The high explosive is connected to a sophisticated triggering device that ensures that all the explosive is detonated at the exact same time. In addition, the array of the high explosive is designed to ensure that all the energy from the explosion is concentrated symmetrically around the sphere of plutonium. This energy compresses the fuel to almost twice its normal density. This means that the ${}^{239}_{94}Pu$ nuclei are much closer together, so that neutrons from the fission of ${}^{239}_{94}Pu$ are

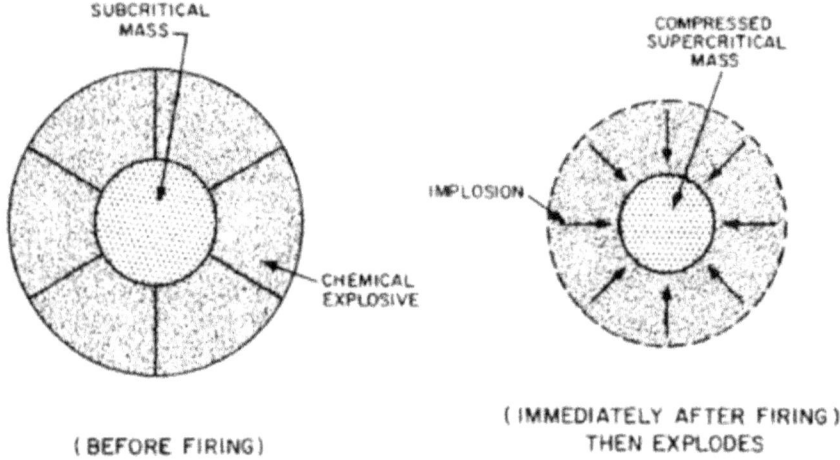

Figure 3.6. Reproduced with permission from [7] (p 16).

not lost but are used to cause additional fissions. In other words, one now has a supercritical mass and a fission explosion occurs. The shock wave produced by the explosion travels at several kilometers per second, while the projectile from the gun-barrel device only travels at a few hundred meters per second [8]. Therefore, the implosion device has a much lower chance of producing a fizzle.

The weapon designers were unsure of the workability of such a device, so it had to be tested before its use over Nagasaki. This design is more technically challenging but requires less nuclear material (3–5 kg). It is the design of choice for those trying to build nuclear weapons that are small enough to be delivered by missile, and it can use either $^{239}_{94}$Pu or $^{235}_{92}$U as the fuel. The US uses implosion devices as part of all its nuclear weapons.

3.11 Proliferation concerns

It should be noted that the same techniques used to enrich uranium for reactor fuel can be used to enrich it so that it contains much higher percentages of $^{235}_{92}$U and can then be used in nuclear weapons. China detonated its first nuclear weapon in 1964, using uranium enrichment technology that was also used to provide fuel for its fledgling nuclear power reactor program. As with enrichment technologies, reprocessing can be used for a nuclear weapons program. India used its reprocessing capabilities to obtain the plutonium to make nuclear weapons, which it demonstrated by exploding a fission weapon in 1975. In addition, one needs to be concerned by the amount of plutonium that is produced in a nuclear reactor and extracted through reprocessing. The plutonium must be safeguarded carefully so that those who wish to make nuclear weapons, or to sell the plutonium to others to make nuclear weapons, are unable to obtain it. The total number of nuclear power reactors worldwide is estimated to produce about 120 tons of plutonium per year. This is enough to form 10 000 critical masses of 10 kg each [6, p 329].

References

[1] www.iaea.org/PRIS/WorldStatistics/OperationalReactorsByCountry.aspx
[2] www.iaea.org/PRIS/WorldStatistics/NuclearShareofElectricityGeneration.aspx
[3] www.iaea.org/PRIS/WorldStatistics/UnderConstructionReactorsByCountry.aspx
[4] Rhodes R and Bellem D 2000 *Foreign Affairs* **70** 30–44
[5] Glasstone S 1982 *The Energy Desk Book* (Oak Ridge, TN: Technical Information Center, US Department of Energy) p 261
[6] Schroeer D 1984 *Science, Technology, and the Nuclear Arms Race* (New York: Wiley) p 31
[7] Glasstone S and Dolan P J 1977 *The Effects of Nuclear Weapons* 3rd edn (United States Department of Defense and the Energy Research and Development Administration)
[8] Hoddeson L, Henriksen P, Meade R and Westfall C 1945 Critical Assembly: *A Technical History of Los Alamos During the Oppenheimer Years 1943-1945* (Cambridge: Cambridge University Press) p 3 (footnote 3)

Chapter 4

The nuclear nonproliferation treaty

After witnessing the incredible destructive power of the nuclear weapons used over Hiroshima and Nagasaki, the international community quickly sensed the need for global security and international arms control. The first attempt to address this issue failed miserably. In 1946 President Truman appointed Bernard Baruch (who also coined the term 'Cold War'), advisor to Presidents Wilson, Hoover and Truman, as US representative to the United Nations Atomic Energy Commission. In 1948 he introduced the so-called 'Baruch plan', in which the United States promised to destroy all its nuclear weapons and nuclear weapons infrastructure in exchange for having all nuclear technology controlled by the United Nations. The Soviet Union promptly rejected the offer.

After the development of nuclear weapons by the Soviet Union, Great Britain, and France, many were concerned about the possibility of other nations following suit. During the famous presidential debate between contenders John Kennedy and Richard Nixon in 1960, Kennedy said 'There are indications because of new inventions, that 10, 15, or 20 nations will have a nuclear capacity, including Red China, by the end of the Presidential office in 1964' [1]. Although he was right about China, the rest of his prediction has fortunately not occurred.

In spite of the difficulty of negotiating a multilateral arms control treaty during the contentious climate of the Cold War, the Geneva Disarmament Conference, consisting of 18 countries (called at that time, the Eighteen Nations Disarmament Committee), was able to successfully conclude the Treaty on the Non-Proliferation of Nuclear Weapons (NPT) in 1968.

The NPT attempts to halt the spread of nuclear weapons by controlling the technologies used to manufacture nuclear weapons. In particular, since the most technically challenging part of developing nuclear weapons is the production of the fuel, the NPT focuses on controlling enrichment and reprocessing capabilities. However, the attempt to control technology has many challenges.

4.1 Why proliferation is a concern

Some have argued that nuclear weapons might actually increase international security. One could argue that a major war between the US and the USSR was averted during the Cold War because both countries had nuclear weapons. If all countries possessed nuclear weapons, a similar deterrent effect would happen and all major wars would be averted. However, there are some problems with that argument.

First of all, countries acquiring nuclear weapons may lack the experience to prevent their use in a time of a crisis. Even the US and the USSR came close to a nuclear exchange during the Cuban Missile Crisis, after possessing nuclear weapons for over 15 years. Most countries that develop nuclear weapons would do so by expending a considerable portion of their resources on such a program, leaving little to invest in command and control technologies. This would make an accident more likely to occur.

There are regions in the world where countries are involved in bitter armed disputes with their neighbors. It is conceivable that a conventional war could escalate to a situation where one of the countries involved might feel an incentive to strike first using nuclear weapons. In addition, some countries have internal problems, and the thought of an unstable regime coming to power and having nuclear weapons is frightening. Finally, the more nuclear weapons there are in the world, the greater the chance is that they or their materials might be stolen by terrorist organizations, or that they might even be given to a terrorist group by a sympathetic government.

Understanding the reasons why a country would want to pursue a nuclear weapons program is extremely helpful in trying to prevent proliferation. Obviously national security is a significant driver. First of all, a country might have a neighbor that has nuclear weapons and believes it would be more secure if it also had them. Or maybe it is surrounded by hostile countries and feels that having a weapon of last resort would enhance its security. Addressing a country's national security concerns through alliances, regional treaties, or other means is extremely important.

There is prestige associated with having nuclear weapons, and a government may want to be part of the exclusive club possessing them. The current members of the nuclear club, defined as those who have exploded a nuclear device, are the United States (1945), Russia (1949), the United Kingdom (1952), France (1960), the People's Republic of China (1964), India (1974), and Pakistan (1998). The date in parentheses denotes the year that they joined the club. Israel is known to have nuclear weapons, but their policy is to neither confirm nor deny that they have them. South Africa was a member of the club, but then got rid of its nuclear weapons and weapons program. The latest member of the club is North Korea, but their first attempts to detonate a nuclear device (2006 and 2009) were less than successful. The yield on the 2006 test was measured to be less that 0.5 kT and the 2009 test showed 2–4 kT [2]. A third test occurred in 2013 and the device exploded had a yield of 6–9 kT, as measured by the Comprehensive Test Ban Treaty Organization, which controls a worldwide array of seismometers [3]. They also detonated a device in January 2016, but it is too early to determine its yield.

Developing a nuclear weapons program provides a country with an advanced technology base. These so-called 'spin-off' technologies can be nuclear power plants, enrichment facilities, or other nuclear-related technologies, such as the use of isotopes for medical purposes. It also provides some autonomy if a country can enrich its own uranium for its nuclear power plants. Addressing these concerns is also very important.

4.2 NPT early history

The following section relies heavily on the NPT fact sheets found on the United States Department of State website for the Bureau of International Security and Nonproliferation[1] and 'talking points' developed by the author in 1994 and 1995.

The increasing arms race between the United States and the Soviet Union, the fear of radiation exposure from atmospheric testing of nuclear weapons, and the increasing number of states that possessed such weapons, caused many in the international community to pursue measures toward reducing and eliminating the nuclear threat. By 1964, there were five declared nuclear weapon states: the United States, the former Soviet Union, the United Kingdom, France, and China. At the same time, there was considerable progress in harnessing the atom for peaceful applications. There were already nuclear reactors operating or under construction in five countries. It rapidly became clear that the spread of nuclear technology would increase the risk of nuclear proliferation and that some credible assurances were needed so that nuclear programs would not be diverted to military applications.

In 1961, the United Nations General Assembly approved a resolution, sponsored by Ireland, calling on all states to conclude an international agreement that would ban the acquisition and transfer of nuclear weapons. In 1965, a conference of nonaligned states (mainly those developing countries that were not part of the Western democracies or the Communist Bloc) held in Cairo developed a series of resolutions that were presented to the UN General Assembly in support of a nonproliferation treaty. In 1965, the Geneva Conference on Disarmament met to discuss the drafts of a nonproliferation treaty submitted by both the United States and the USSR.

The major problem in reaching an agreement had to do with the NATO alliance. The US wanted to distribute nuclear weapons among the NATO allies. The Soviets were opposed to this, especially when it came to those units stationed in then West Germany. Under pressure from the other countries at the conference, the United States and the Soviet Union reached agreement on the text of a treaty in 1967, after US assurances that they would maintain exclusive control of their nuclear weapons stationed in NATO countries.

The negotiations at Geneva were completed in 1968 following consultations with other states. On 1 July 1968, the NPT was opened for signature, and it entered into force on 5 March 1970, with 45 members. The three nuclear-weapons states (NWS) that originally signed the NPT (the United States, the United Kingdom, and the USSR) were designated the 'depositary governments' and were responsible for all administrative duties associated with the treaty.

---

[1] www.state.gov/t/isn/trty/16281.htm

4.3 Conference on Disarmament

The Conference on Disarmament (CD) is still in existence. Although originally an informal organization, it was established in 1979 as the single multilateral disarmament negotiating forum of the international community as a result of the first Special Session on Disarmament of the United Nations General Assembly held in 1978. There are currently 65 member states: Algeria, Argentina, Australia, Austria, Bangladesh, Belarus, Belgium, Brazil, Bulgaria, Cameroon, Canada, Chile, China, Colombia, Cuba, Democratic People's Republic of Korea, Democratic Republic of the Congo, Ecuador, Egypt, Ethiopia, Finland, France, Germany, Hungary, India, Indonesia, Iran, Iraq, Ireland, Israel, Italy, Japan, Kazakhstan, Kenya, Malaysia, Mexico, Mongolia, Morocco, Myanmar, the Netherlands, New Zealand, Nigeria, Norway, Pakistan, Peru, Poland, Republic of Korea, Romania, Russian Federation, Senegal, Slovakia, South Africa, Spain, Sri Lanka, Sweden, Switzerland, Syria, Tunisia, Turkey, Ukraine, the United Kingdom, the United States, Venezuela, Vietnam, and Zimbabwe.

They successfully negotiated the following international treaties: NPT, the Convention on the Prohibition of Military or Any Other Hostile Use of Environmental Modification Techniques, the Seabed treaties, the Convention on the Prohibition of the Development, Production and Stockpiling of Biological Weapons, the Convention on the Prohibition of the Development, Production, Stockpiling and Use of Chemical Weapons, and the Comprehensive Nuclear-Test-Ban Treaty.

There are still many important items that need to be addressed by the CD. However, as with many international conferences, all decisions must be made by consensus. There has been no agreement on even an agenda for continuing negotiations for the last 18 years. Although China was responsible for blocking the agenda early on over concerns about the weaponization of space, Pakistan has been blocking consensus now for the past 10 years because it does not want discussion on a global ban on the future production of fissile materials (highly enriched uranium and plutonium) for nuclear weapons (the Fissile Material Cutoff Treaty, or FMCT) unless existing stockpiles of fissile material are eliminated.

Many negotiators are frustrated, as evidenced by the following quotation: 'while the international community has been active and achieved results in many areas during the past year, the Conference on Disarmament appears to be no closer to an 'honest day's work' than it was last January' [4]. For the third year in a row, an informal working group could not reach consensus on any of the CD's four core issues: nuclear disarmament, the FMCT, the prevention of an arms race in outer space, and negative security assurances (nuclear-weapons states promising not to use their nuclear weapons against states that do not possess them).

4.4 The NPT

The complete text of the NPT is shown in the appendix. It is short in length, but long on substance. A summary of the treaty follows.

The NPT recognized the fact that there were five nuclear-weapons states at the time the treaty went into force. All other parties to the NPT were then recognized as non-nuclear weapons states (NNWS). In fact, any country joining the NPT must do so as a NNWS.

The NPT's primary purpose is to prevent the future spread of nuclear weapons. Article I of the treaty prohibits NWS from sharing nuclear weapons and nuclear weapons technology with NNWS. Article II prohibits NNWS from acquiring, by any means, nuclear weapons or nuclear weapons technology. Article III of the NPT provides assurances, through international safeguards inspections, conducted by the International Atomic Energy Agency (IAEA), that the peaceful nuclear activities of NNWS will not be diverted to the making of nuclear weapons.

The NPT provides a reward for the NNWS for abiding by the terms of the first three articles. It promises to promote, to the maximum extent possible, and consistent with the purposes of the treaty, the peaceful uses of nuclear energy, especially in developing countries. It seeks to create confidence that certain types of cooperation in the peaceful uses of nuclear energy can occur without creating an unacceptable risk of nuclear proliferation.

Those NPT member states with large civil nuclear infrastructures have engaged in extensive nuclear export programs over the years, with preference shown to NPT parties. The NPT has helped to promote nuclear energy projects in Mexico, Argentina, Brazil, and South Africa. Furthermore, the IAEA, in addition to its safeguards duties, has an extensive technical cooperation program designed to assist smaller countries with non-energy applications. The IAEA has introduced nuclear medicine into more than 40 developing countries. Nuclear techniques have assisted efforts to increase crop production, manage water resources, and help eradicate destructive insects. There are extensive programs in developing countries to improve the safe handling and disposition of small quantities of radioactive sources.

The developing countries that are party to the NPT view this international cooperation as the major benefit that they derive from the NPT. As the many benefits of the NPT became clear to countries around the world, NPT membership grew steadily over the years. Japan, South Korea, and the then five non-nuclear-weapon states of the European Atomic Energy Community joined the NPT in the mid 1970s. There were also other significant milestones. Egypt joined in 1981 and South Africa in 1991. China and France, the two remaining nuclear weapon state holdouts, joined in 1992, as did all the states of the former Soviet Union by 1994. Argentina's accession in 1995, Brazil's in 1998, and Cuba's in 2002 brought membership up to today's total of 186.

Article V of the NPT mentions the benefits of peaceful nuclear explosions. This may seem like a contradiction in terms, but at least for a while, some believed nuclear weapons could be used for major construction efforts. A nuclear weapon, exploded underground, vaporizes the Earth around it. The soil above then collapses inward, forming a large cavity. Several nuclear weapons, exploded in sequence along a straight line, can create a large trench. The USSR explored this by trying to alter the flow of some rivers in Siberia for irrigation purposes, but stopped this practice because of the radioactive contamination. The US Army Corps of Engineers helped to develop a plan to use nuclear explosions to dig a second canal in Central America, in either Panama or Nicaragua. It was part of a series of 35 peaceful nuclear explosion tests called Project Plowshare that took place from 1958 through 1975. The US finally decided in 1970 that the use of nuclear explosions to dig a new Panama Canal was not feasible.

Article VI of the treaty is the most contentious. Although it contains no explicit requirement for the five NWS, it does obligate them to engage in good faith negotiations to end the nuclear arms race and to progress towards nuclear disarmament. It also requires all members to work towards a treaty on total disarmament under strict international inspections. Most members of the NPT believe that the NWS have not done enough to fulfill their Article VI obligations. However, there have been some dramatic changes.

Since the fall of the Berlin Wall, the United States and the former Soviet Union (now the Russian Federation) have taken many dramatic steps to reduce Cold War stockpiles of nuclear weapons. They have eliminated all intermediate-range and short-range missiles, and have dismantled most of their associated warheads. In all, tens of thousands of nuclear weapons have been dismantled. They have pledged to reduce nuclear weapons even further, as evidenced by the so-called 'New START' treaty. They have also addressed the security and disposition issues related to nuclear material formerly used in military applications. The US and Russia have removed hundreds of tons of nuclear material from military stockpiles, and the two countries have cooperated in efforts to render this material unusable for nuclear weapons. Of the five NWS, all but China have publicly declared that they are not producing any more nuclear material for nuclear weapons and, until recently, all five have supported the negotiation of a global treaty permanently banning such production. The United Kingdom and France have additionally taken many unilateral steps to cut back their nuclear weapons programs.

The United States and Russia signed the Comprehensive Test Ban Treaty (CTBT), although the US has not ratified it. Even though the CTBT has not yet entered into force, all the NWS are observing moratoria on the testing of nuclear explosives. The US last tested in 1992.

The NPT required that a conference be held in 1975 to review the operation of the NPT. Such review conferences (RevCons) could be held every five years thereafter if a majority of parties agreed to do so. As result, nine NPT RevCons have met. They take place over a four-week period, with at least three preparatory conferences prior to the main meeting. The participants divide themselves into three committees, one dealing with nuclear proliferation and arms control issues, one with nuclear safeguards, and one with nuclear cooperation. Each committee attempts to produce a report that then goes to the plenary for a final document. These documents not only assess the current status of the treaty, but also provide guidance on what actions in the future might strengthen it. Before each RevCon there are three preparatory committee (PrepCom) meetings to decide on the chairs for each of the three committees and the agenda for the RevCon.

The final document must be agreed to by consensus, resulting in a report that encompasses a wide range of important policy issues. Due to the wide divergence of views, particularly on the pace of nuclear disarmament and access to nuclear technology, only the 1975, 1985, 2000, and 2010 RevCons have been able to reach agreement on a final document.

The NPT was originally designed to be in force for only 25 years. Some countries, particularly Italy, did not want to forgo a nuclear weapons option when the treaty went into force in 1970. Those countries wanted to see how the NPT operated in the intervening time. Article X.2 of the NPT required that after 25 years the treaty parties meet to decide its future. The article said that the decision would be reached by a vote, not consensus, and it gave the parties three options. The treaty could either be extended indefinitely (i.e. made permanent), it could end in 1995, or it could be 'extended for an additional fixed period or periods'.

4.5 Recent history of the NPT

In 1995, the parties met in New York at the United Nations for another five-year RevCon and an extension conference. It was a major test of the treaty's viability. The US strategy, formulated by the Arms Control and Disarmament Agency (ACDA)[2], was to achieve the indefinite extension of the NPT, by a majority of one vote if necessary. The ACDA was established as an independent agency to ensure that arms control was fully integrated into the development and conduct of United States national security policy. It also conducted, supported, and coordinated research for arms control and disarmament policy formulation, prepared for and managed US participation in international arms control and disarmament negotiations, and prepared, operated, and directed US participation in international arms control and disarmament systems.

It was clear to many that the NWS had not done enough towards eliminating their nuclear weapons, and some NNWS had decided not to support an indefinite extension of the treaty. Some proposals called for 'rolling' extensions, which meant that every 10 or 25 years the parties would meet to consider whether or not to continue the NPT. The US view was that indefinite extension would create a predictable environment for future arms control measures.

After intense lobbying by the US and other supporters of indefinite extension, a majority existed in support of the US position. In fact, a vote was never taken. The final document, passed by consensus, stated that a consensus existed that a majority supported the indefinite extension of the NPT. Over 120 countries, of the 178 that were party to the treaty at that time, either publicly or privately supported the indefinite extension of the NPT.

Along with the extension consensus, there were decisions on strengthening the NPT review process, and on principles and objectives for nuclear nonproliferation and disarmament. The latter decision set forth a number of elements on which progress was expected by the time of the 2000 RevCon, including the conclusion of a CTBT. Finally, a special statement on the Middle East, proposed by Egypt, called for all states in the region to accede to the NPT. An interesting note is that even though the parties agreed on the indefinite extension of the NPT, they could not reach consensus on a final document.

---

[2] ACDA was disbanded in 1997 and its duties folded into the State Department in a deal between President Clinton and Senator Jesse Helms of North Carolina in order to get the Chemical Weapons Convention Treaty, outlawing chemical weapons, ratified by the US Senate.

The next five years before the 2000 RevCon were difficult ones for the NPT. North Korea and Iraq were found to have violated their NPT agreements. India and Pakistan tested nuclear weapons in 1998. The Middle East was still a problem, and there was growing criticism of Israel's nuclear weapons from Arab countries. The US and the other NWS disagreed over the future of the Anti-Ballistic Missile (ABM) Treaty. The CTBT had been concluded in 1996, but US Senate failed to muster the votes to ratify it in the fall of 1999.

However, even considering all these challenges to the NPT, a final document was agreed upon. The NWS were willing to ignore their differences for the sake of the NPT. The US and Egypt had arrived at a balanced approach to the Middle East prior to the conference. The document was essentially a continuation of the goals outlined in the 1995 review conference final document to strengthen the treaty review process and pursue specific principles and objectives for nuclear nonproliferation and disarmament.

A coalition of NNWS, called the New Agenda Coalition (NAC), was launched in 1998 by the foreign ministers of eight nations—Ireland, Brazil, Egypt, Mexico, New Zealand, South Africa, Slovenia, and Sweden—with the purpose of pressuring the NWS to fulfill their obligations from Article VI of the NPT. They worked constructively with the NWS to produce a realistic approach to Article VI issues, called the '13 steps', named after the 13 steps contained in the final document. These steps are listed below [5].

1. **Comprehensive nuclear test ban treaty**
   The importance and urgency of signature and ratification, without delay and without conditions and in accordance with constitutional processes, to achieve the early entry into force of the Comprehensive Nuclear Test Ban Treaty.

2. **Nuclear test moratorium**
   A moratorium on nuclear weapon test explosions or any other nuclear explosions pending entry into force of that Treaty.

3. **Fissile material cutoff treaty**
   The necessity of negotiations in the Conference on Disarmament on a non-discriminatory, multilateral and internationally and effectively verifiable treaty banning the production of fissile material for nuclear weapons or other nuclear explosive devices in accordance with the statement of the Special Coordinator in 1995 and the mandate contained therein, taking into consideration both nuclear disarmament and nuclear non-proliferation objectives. The Conference on Disarmament is urged to agree on a program of work, which includes the immediate commencement of negotiations on such a treaty with a view to their conclusion within five years.

4. **Nuclear disarmament discussions**
   The necessity of establishing in the Conference on Disarmament an appropriate subsidiary body with a mandate to deal with nuclear disarmament. The Conference on Disarmament is urged to agree on a program of work which includes the immediate establishment of such a body.

5. **Irreversibility of nuclear reductions**
   The principle of irreversibility to apply to nuclear disarmament, nuclear and other related arms control and reduction measures.

6. **Elimination of nuclear arsenals**
   An unequivocal undertaking by the nuclear-weapon States to accomplish the total elimination of their nuclear arsenals leading to nuclear disarmament to which all State parties are committed under Article VI.

7. **The START II, START III, and ABM treaties**
   The early entry into force and full implementation of START II and the conclusion of START III as soon as possible, while preserving and strengthening the ABM Treaty as a cornerstone of strategic stability and as a basis for further reductions of strategic offensive weapons, in accordance with its provisions.

8. **Securing excess nuclear material**
   The completion and implementation of the Trilateral Initiative between the United States of America, the Russian Federation, and the International Atomic Energy Agency.

9. **Other nuclear-weapon states' actions**
   Steps by all the nuclear-weapon States leading to nuclear disarmament in a way that promotes international stability, and based on the principle of undiminished security for all:
   - *Further efforts by the nuclear-weapon States to reduce their nuclear arsenals unilaterally*
   - *Increased transparency by the nuclear-weapon States with regard to the nuclear weapons capabilities and the implementation of agreements pursuant to Article VI, and as a voluntary confidence-building measure to support further progress on nuclear disarmament*
   - *The further reduction of non-strategic nuclear weapons, based on unilateral initiatives, and as an integral part of the nuclear arms reduction and disarmament process*
   - *Concrete, agreed measures to further reduce the operational status of nuclear weapons systems*
   - *A diminishing role for nuclear weapons in security policies to minimize the risk that these weapons ever be used, and to facilitate the process of their total elimination*

10. **Excess fissile material**
    The engagement, as soon as appropriate, of all the nuclear arrangements by all nuclear-weapon States to place, as soon as practicable, fissile material designated by each of them as no longer required for military purposes under IAEA or other relevant international verification, and arrangements for the disposition of such material for peaceful purposes, to ensure that such material remains permanently outside of military programs.

11. **General and complete disarmament**
    Reaffirmation that the ultimate objective of the efforts of States in the disarmament process is general and complete disarmament under effective international control.

12. **Regular reports on disarmament progress**
    Regular reports, within the framework of the NPT-strengthened review process, by all States parties on the implementation of Article VI and paragraph 4 (c) of the 1995 Decision on "Principles and Objectives for Nuclear Non-Proliferation and Disarmament," and recalling the Advisory Opinion of the International Court of Justice of 8 July 1996.

13. **Verification**
    The further development of the verification capabilities that will be required to provide assurance of compliance with nuclear disarmament agreements for the achievement and maintenance of a nuclear-weapon-free world.

Unlike the 2000 review conference, the 2005 RevCon ended in failure. President George W Bush was now in office and with the new US administration's actions (abrogating the ABM, abandoning the START process, and NMD deployment) it became clear that most of the 13-step approach to disarmament would never be achieved. A compromise position was drafted and presented at the 2005 RevCon by a newer version of the NAC. The coalition now consisted of seven non-nuclear nations: Ireland, Brazil, Egypt, Mexico, New Zealand, South Africa, and Sweden.

The compromise approach would allow for states/parties to focus their attention on reaching agreements on the nuclear disarmament obligations, commitments, and undertakings that could be implemented in the foreseeable future and in the period before 2010, without negating those that were agreed on in 1995 and 2000. Some of the more specific measures were as follows [6].

- *The necessity to achieve the early entry into force of the CTBT while maintaining existing moratoria on nuclear testing;*
- *The need for the nuclear-weapon states to take further steps to reduce their nonstrategic nuclear arsenals, and not to develop new types of nuclear weapons in accordance with their commitment to diminish the role of nuclear weapons in their security policies;*
- *The need for the Conference on Disarmament (CD) urgently to resume negotiations on an internationally and effectively verifiable treaty banning the production of fissile material for nuclear weapons or other nuclear explosive devices, taking into account both nuclear disarmament and nuclear nonproliferation objectives;*
- *The completion and implementation of arrangements by all nuclear-weapon states to place fissile material no longer required for military purposes under international verification;*
- *The establishment of an appropriate CD subsidiary body to deal with nuclear disarmament;*

- *The imperative of the principles of irreversibility and transparency for all nuclear disarmament measures, and the need further to develop adequate and efficient verification capabilities.*

However, the US declared that the NAC's original 13-step approach, agreed to by the US in 2000, was no longer valid. The US delegation also refused to consider the NAC's compromise proposal. The US delegation claimed that the world had changed after the 11 September 2001 attacks on the US, and that US and international security from terrorism required a new approach. The US preferred to concentrate on the North Korean and Iranian situations. This approach angered many NPT parties, and no final document was ever drafted, let alone approved.

Many NNWS were then very concerned about the future viability of the NPT. The process for the 2010 RevCon did not begin well. The first PrepCom meeting, in May 2007, ended without the approval of a final document, mostly because of the language used to address the Iranian situation. The second PrepCom was not even able to decide on the agenda for the 2010 RevCon. However, with the new Obama administration in place and a new outlook on nuclear nonproliferation from the US, the third and final PrepCom ended on a positive note.

The 2010 RevCon was an important test of the viability of the NPT. Actions by the Obama administration positively affected the outcome of the NPT conference. Negative security assurances (stating that NWS would not use nuclear weapons against NNWS) are important to the NNWS and the Nuclear Posture Review of 2010 reaffirmed the US commitment not to use nuclear weapons against those NNWS who are party to the NPT and are living up to their nonproliferation obligations. The New START treaty also demonstrated the US commitment to reducing its nuclear arsenal with a verifiable treaty. All NPT parties viewed President Obama's speech in Prague envisioning a world without nuclear weapons quite favorably. In addition, the US commitment to seek ratification of the Comprehensive Test Ban Treaty and to begin negotiations on a Fissile Material Cutoff Treaty had a positive impact.

After considerable negotiations and last-minute intrigue, the NPT parties agreed, by consensus, to a final document. Although lacking in precise details, the final document did affirm the parties' commitment to the three pillars of the NPT: disarmament, safeguards, and nuclear cooperation. At the last minute, Egypt threatened to break the consensus unless there was stronger language concerning Israel. The NPT parties agreed to hold a conference in 2012 concerning the establishment of a weapons-of-mass-destruction (WMD) free zone in the Middle East.

That conference did not happen and concerns about the Middle East derailed any hope of a consensus final document at the 2015 RevCon. Wording in the draft final document stated that the conference must be held by March 2016 and that all parties affected would be invited [7]. If one or more parties did not attend, the conference would still be held. The United States, the United Kingdom, and Canada blocked consensus because of concerns regarding setting a certain date and not requiring all parties to be present.

Other strong concerns were raised by the NNWS over the slow pace of disarmament by the United States and Russia. There have been no significant reductions by either country since the New START treaty. The deteriorating state of US/Russian relations and recent plans by both countries to modernize their nuclear forces indicated to the NNWS that neither country was taking their Article VI commitments seriously. Although the impasse on the Middle East is credited for being the issue that blocked consensus, the widespread dissatisfaction with the lack of progress on nuclear disarmament points to a more serious situation.

4.6 The IAEA, the Nuclear Suppliers Group and the Zangger Committee

The IAEA is intimately tied to the NPT, although supporting the treaty was not its original mission. There are another two organizations, the Zangger Committee and the Nuclear Suppliers Group (NSG), that help support the goal of nuclear nonproliferation by attempting to ensure nuclear-related exports are not misused.

4.6.1 The IAEA

In 1953, President Eisenhower proposed his Atoms for Peace Program. It was designed to promote the peaceful use of nuclear energy worldwide, while calming some of the fears of the US population concerning nuclear war. As part of his announcement, he proposed the establishment of an independent international organization to deal with nuclear energy issues. The IAEA was founded by statute in 1957. It is an autonomous agency of the United Nations, and its headquarters is in Vienna, Austria. The IAEA is composed of 138 member states. It has a director general, a secretariat, and a 35-person board of governors. There are 10 permanent members of the board, representing the 10 most industrially advanced countries, and 25 other rotating members, chosen to represent the world geographically.

The IAEA mission statement, found on their website [8][3], says the following:
'The ... IAEA:
- is an independent intergovernmental, science and technology-based organization, in the United Nations family, that serves as the global focal point for nuclear cooperation
- assists its Member States, in the context of social and economic goals, in planning for and using nuclear science and technology for various peaceful purposes, including the generation of electricity, and facilitates the transfer of such technology and knowledge in a sustainable manner to developing Member States
- develops nuclear safety standards and, based on these standards, promotes the achievement and maintenance of high levels of safety in applications of nuclear energy, as well as the protection of human health and the environment against ionizing radiation
- verifies through its inspection system that States comply with their commitments under the Non-Proliferation Treaty and other non-proliferation agreements, to use nuclear material and facilities only for peaceful purposes.'

---
[3] www.iaea.org/worldatom/About/Profile/mission.html.

The IAEA is considered to be the world's expert on radiation safety. The IAEA also investigates nuclear and radiation accidents around the world. However, we are most interested in the IAEA's duties associated with the NPT.

As important as safeguards inspections are, they only are a part of the IAEA's mission. The IAEA has a staff of about 2500 people, of whom only 850 are in the Safeguards Division, including inspectors, researchers, and administrative staff. Approximately half of them conduct inspections. The total budget for 2014 was about 370 million US dollars (USD), while only 140 million USD went to the Safeguards Department. Member states are charged a membership fee that is calculated according to their economic situation, as is done in the UN. In addition, member states make voluntary contributions to fund safeguards research or technology assistance programs. The US provides over 60 million USD, plus an almost equal amount in voluntary extra-budgetary assistance.

The IAEA is responsible for the safeguards agreements required from all NNWS under the NPT. Each NNWS is required to conclude safeguards agreements with the IAEA once they ratify the treaty. The original safeguards agreement, outlined in Information Circular (INFCIRC) 153, required that a state declare its nuclear facilities and nuclear material. The IAEA then verifies, through monitoring records, the use of tamper-proof seals, remote monitoring and sensors, on-site inspections, and any combination of these methods, that nuclear material is not being diverted. Although they cannot detect the diversion of a small amount of material, they are required to verify that a significant diversion has not taken place and to do this in a timely manner to ensure that the diverted material is not used to build a weapon.

The IAEA safeguards over 1200 facilities in 180 different countries. These facilities include power and research reactors, enrichment and reprocessing facilities, fuel fabrication plants, and storage facilities. They maintain accountancy over 800 metric tons of plutonium, 20 metric tons of HEU, and 55 000 metric tons of low-enriched uranium. The IAEA does this with a staff of about 225 inspectors, who conduct approximately 2800 inspections (over 111 000 calendar days in the field) annually [9].

Most countries were satisfied with the inspection regime until the first Gulf War. At that time, it was discovered that the Iraqis had a clandestine nuclear weapons program. The problem with INFCIR 153 was that it authorized the IAEA to safeguard a country's declared facilities. The IAEA had no internal intelligence apparatus. Nor did it want to be seen as receiving intelligence information from either Western countries or the Communist bloc. It could only rely on what a country declared and did not have the authority to search throughout a country.

The IAEA responded with their enhanced safeguards protocol. They outlined additional safeguards requirements in INFCIRC 540; a country is considered to have an enhanced safeguards arrangement if it has a full scope safeguards agreement (INFCIRC 153) plus the additional protocol. Under this arrangement, a country prepares a list of its facilities and materials that will require safeguards. The IAEA, with help from member states and their own intelligence assets, prepares a similar list. The IAEA then works with a country to try to handle any discrepancies.

The IAEA then have the right to inspect anywhere in a country to ensure that no material is being diverted from peaceful purposes. At the same time, to be able to handle the extra work with its existing staff, the IAEA cannot inspect every facility within a country every year. A random procedure, weighted by the importance of the facilities, is used. In addition, during an announced inspection, an IAEA inspector has the right to immediately inspect any building on the site of the facility being inspected.

4.6.2 The NSG

The NSG was formed in 1974 following the Indian nuclear explosion test, deemed a peaceful nuclear test by the Indian government. It clearly demonstrated to the world how the technology associated with a civilian nuclear power program could be diverted from peaceful purposes toward building nuclear weapons. The NSG's initial purpose was to develop guidelines to control the flow of nuclear technology and to help prevent its misuse.

The first set of guidelines was called the 'trigger list'. It was a list of equipment and material that, if exported, would trigger IAEA safeguards inspections. In order for the material or equipment to be exported, the country receiving the equipment would need to have a safeguards agreement with the IAEA for the facility receiving the exported items. The second set of guidelines listed so-called 'dual-use' equipment and material. These dual-use items have an original purpose that is not weapons related, but they can be applied to a weapons program. The NSG guidelines were issued in 1974 and members were to then conform voluntarily to the guidelines. They were designed to establish a requirement for formal notification from countries in receipt of listed items that there were IAEA safeguards agreements in place, or that that there was adequate protection against the material and equipment being be diverted to weapons programs. The guidelines also urged NSG members to exhibit caution when exporting dual-use items.

The NSG meets annually in plenary meetings and through special working groups throughout the year, to enhance nuclear nonproliferation worldwide. In that regard they also work closely with the IAEA and participate in all NPT review conferences. There have been many additions to the dual-use and trigger lists over the years, which must be agreed to by consensus. After the first Gulf War in 1991 and the discovery of the Iraqi nuclear weapons program in spite of IAEA safeguards inspections, the NSG reacted by amending their guidelines. A country receiving trigger list items must have a full-scope safeguards agreement with the IAEA that covers the whole country, not just the facility receiving the material. They also strengthened their guidelines for the export of dual-use items.

There are currently 45 members of the NSG (detailed below). The current issues being discussed within the NSG involve adding items to the trigger list or possibly requiring full-scope safeguards for even dual-use items. There is also an effort to reach out to nonmembers such as Egypt, India, and Iran.

The current NSG members are: Argentina, Australia, Austria, Belarus, Belgium, Brazil, Bulgaria, Canada, China, Croatia, Cyprus, Czech Republic, Denmark, Estonia, Finland, France, Germany, Greece, Hungary, Iceland, Ireland, Italy, Japan, Kazakhstan, Republic of Korea, Latvia, Lithuania, Luxembourg, Malta, the Netherlands, New Zealand, Norway, Poland, Portugal, Romania, Russia, Slovakia, Slovenia, South Africa, Spain, Sweden, Switzerland, Turkey, Ukraine, the United Kingdom, and the United States.

4.6.3 Zangger Committee

The Zangger Committee is similar to the NSG in that it also establishes guidelines for the export of nuclear-related items and material. It is named after Professor Claude Zangger of Switzerland. He chaired a series of informal meetings from 1971 to 1974 among 15 nuclear supplier states, mostly members of the NPT, to discuss implementation of Article III of the NPT. Article III requires that NNWS have all their nuclear technology and material under IAES safeguards. In that regard, they also adopted a trigger list of items that would require IAEA safeguards, and arrived at a consensus on the 'rules of the game' with regards to exports and NPT obligations.

There are some important differences between the Zangger Committee, consisting of 36 member states, and the NSG. First of all, the Zangger Committee works within the NPT. Their trigger lists and other policies are submitted to the IAEA and published in IAEA information circulars. Their structure is more informal than the NSG's; therefore, they are able to be more responsive to current world situations, and are able to take the lead on a number of important NPT and nuclear proliferation matters.

The current Zangger Committee members are: Argentina, Australia, Austria, Belgium, Bulgaria, Canada, China, Croatia, Czech Republic, Denmark, Finland, France, Germany, Greece, Hungary, Ireland, Italy, Japan, Kazakhstan, Republic of Korea, Luxemburg, the Netherlands, Norway, Poland, Portugal, Romania, Russia, Slovakia, Slovenia, South Africa, Spain, Sweden, Switzerland, Turkey, Ukraine, the United Kingdom and the United States.

4.7 Successes of the NPT

The combination of the NPT, the IAEA, and the export control regime has provided a stable platform to support the global nuclear nonproliferation efforts envisioned by the NPT. Although there are challenging times ahead, the NPT remains strong. It is the principal legal and political barrier against proliferation and, because of its global scope, reflects an international norm of nonproliferation. Some of its major successes over its 45 year existence are outlined below.

South Africa dismantled its nuclear weapons prior to its accession to the NPT in 1991. It is the only country ever to have done so. All its nuclear facilities and nuclear material were subjected to verification by the IAEA. These steps enhanced the security of all African states, and opened the way to the achievement of the African Nuclear Weapons-Free Zone Treaty, which bans nuclear weapons, either indigenous or foreign, from the entire African continent.

During the 1980s there was concern about a possible a nuclear arms race occurring in Latin America, particularly among Argentina, Chile, and Brazil. Although there was a nuclear weapons free zone treaty in Latin America (the Treaty of Tlatelolco), it never entered into force. In the early 1990s, Argentina and

Chile (with encouragement from the United States) brought the treaty into force, obligating them not to develop nuclear weapons. Argentina and Brazil also adopted a safeguards agreement with the IAEA. Finally, Argentina led the way by acceding to the NPT in 1995.

In Europe and Central Asia, the dissolution of the former Soviet Union and the Warsaw Pact created enormous turbulence, but the NPT provided stability. Kazakhstan, Belarus, and Ukraine—former states of the Soviet Union with nuclear weapons stationed in their territory—transferred all their nuclear weapons to Russia and joined the NPT as NNWS. Peace between Israel and Egypt in the late 1970s was soon followed by Egypt's accession to the NPT. In the Middle East, only Israel has not joined the NPT. The NPT has also convinced countries in Southeast Asia to remain nuclear free. However, in terms of the future of international nonproliferation efforts, the most important success associated with the NPT was its indefinite extension.

4.8 Difficulties faced by the NPT

In his State of the Union Address in January 2002, President George W Bush referred to Iran, Iraq, and North Korea, stating that '(s)tates like these, and their terrorist allies, constitute an axis of evil, arming to threaten the peace of the world' [10]. Although he was talking about each state in a broader context, he was also concerned about their nuclear weapons ambitions. These three countries, along with Libya and Syria, have presented particular challenges for the NPT. In fact, the advanced state of their efforts surprised many, including the US government and the IAEA.

4.8.1 Iraq

Had Saddam Hussein not miscalculated regarding the US response to his invasion of Kuwait, the world might have not known about his nuclear weapons program until it was too late to stop it. Iraq had taken advantage of the original IAEA Safeguards Agreement, which only permitted the IAEA to inspect nuclear facilities and material that had been declared by the state. Clearly Iraq left much undeclared. This situation led the IAEA to develop its enhanced safeguards protocols.

Iraq was one of the original signatories of the NPT. They received some help with nuclear technology through the Atoms for Peace Program in the 1950s. By 1970, responding to the threat posed by Israel's nuclear program, Iraq began a clandestine nuclear weapons effort. In 1976, the Iraqis purchased a 40 MW(th) reactor from the French called 'Osirak'. Just before it was loaded with fuel in 1981, the Israeli air force attacked and destroyed the reactor over concerns about Iraq's nuclear weapons program. This act reinforced Iraq's interest in developing weapons and accelerated their efforts.

By the time the first Gulf War ended in 1991, it was clear that the Iraqis had a mature nuclear weapons program. They had enriched uranium purchased from France and Russia for their Osirak reactor and were working on their own enrichment program. They were using calutrons (short for California University cyclotron, a technique developed and then abandoned by the Manhattan Project as being too inefficient) and centrifuges. They had a design for an implosion device that used highly enriched uranium as the fuel. They also had high-explosive material and the appropriate electronics to assemble their device.

Although there were many details that needed to be addressed, experts think that they were anywhere from six months to two years away from having a working weapon. Their program was completely destroyed during and after the first Gulf War and, contrary to what was stated by President George W Bush's administration before the invasion of Iraq, they never restarted their efforts.

4.8.2 Libya

Libya started exploring the nuclear weapons option in the 1970s, when they asked China if they could purchase a weapon. However, by all outward appearances, they were demonstrating their support for the NPT. They joined the treaty in 1975 and supported indefinite extension in 1995. They also signed the African Nuclear Weapons-Free Zone Treaty in 1996.

Because of Libya's support for terrorist organizations and its involvement with those responsible for the downing of a US passenger jet over Lockerbie, UK in 1988, the UN and individual countries such as the US imposed strict sanctions in the late 1980s and early 1990s. These sanctions were very effective and in 2003 Libya, in an attempt to have the sanctions lifted, renounced terrorism and divulged its clandestine nuclear program to the IAEA.

Libya had conducted unreported research on enrichment and reprocessing. They had centrifuges and an old Chinese weapons design (obtained from A Q Khan[4]). The IAEA was able to verify all the Libyan claims and provided the international community with assurances that Libya no longer had a weapons program. The US and UN lifted all sanctions and Libya received US assistance to get its economy back on track until the 'Arab Spring' and the ongoing civil war.

4.8.3 Syria

Syria is an even more interesting case. The Syrians made several attempts to obtain a research reactor. In 1984, they worked with the IAEA on purchasing a research reactor, but negotiations broke down. They tried to buy a reactor from China or Argentina in the early 1990s, and Russia during the late 1990s and early 2000s, but nothing ever came of these efforts because Syria did not have the resources.

Then in September 2007, the Israelis launched an airstrike against a target inside Syria. The US and other countries claimed that the site contained a building, similar

---

[4] Abdul Qadeer Khan was dubbed the father of Pakistan's nuclear weapons program. He began Pakistan's enrichment program with plans and equipment he smuggled from the Netherlands in 1975. In 2004, it was discovered that Khan had an extensive black-market enterprise that sold centrifuges, centrifuge plans, and even nuclear weapons plans to countries such as Libya, Iran, and North Korea. He had shadow companies set up around the globe to support his efforts. Some have suggested that Pakistan encouraged him, or at least, being aware of what was happening, did nothing to impede him. Pakistan vehemently denies this. The full extent of his enterprises is still not known. After five years of house arrest, Pakistan declared him a free man in 2009.

in size and shape to the one that housed the North Korean 5 MW(e) research reactor. There were also photos and videos released that purported to show North Korean nuclear technicians at the site before the bombing.

After the bombing, the Syrians worked extremely quickly to clean up the site. The IAEA asked to inspect it and was initially granted permission. When their first inspection showed evidence of suspicious activity, the IAEA asked to go back and make a second inspection in August 2008, but the Syrians refused to allow this. Neither Syria nor North Korea ever commented on the matter further and the US and the IAEA continued to pressure the Syrians for further inspections. Of course these are now on hold because of the current civil war.

4.8.4 DPRK

North Korea, or as it is officially known, the Democratic Peoples Republic of Korea (DPRK), began its nuclear aspirations in the 1960s. In 1979, they began construction of a 5 MW(e) (30 MW(th)) research reactor. After considerable pressure on the Soviet Union by the United States and other countries, the Soviets were able to convince the North Koreans to accede to the Nuclear Nonproliferation Treaty in 1985. However, due to an administrative error on the part of the IAEA, which the North Koreans exploited, and other North Korean demands, the DPRK did not conclude their Safeguards Agreement with the IAEA until 1992.

Between 1985 and 1992, the North Koreans worked diligently on their nuclear program. They started construction of a 200 MW(th) reactor and a plutonium reprocessing facility in 1985. They began operating their smaller 5 MW(e) reactor in 1986. In 1989, they unexpectedly shut down this reactor for 70 days. Some (if not all) of the exposed fuel rods were removed and some plutonium was reprocessed, although the North Koreans have not fully disclosed details about what actually happened. Also in 1989, the DPRK began work on an even larger 800 MW(e) reactor at Yongbyon and in 1990, according to US intelligence agencies, they began constructing what was determined to be a high-explosive test facility.

By the time the IAEA's first safeguards inspections occurred, most of the international community suspected that the DPRK had a mature nuclear weapons program in place. Although the North Koreans claimed that they had only experimented with reprocessing on one occasion, the IAEA was able to determine, through careful analysis, that reprocessing had occurred on at least three different dates over a two-year period. When the IAEA requested access to other DPRK facilities, the government refused and ended IAEA inspections. The IAEA then informed the United Nations Security Council that the DPRK was in violation of its Safeguards Agreement.

As the UN Security Council discussed possible sanctions against the DPRK for their activities, the DPRK became quite bellicose. They started moving large numbers of troops toward the border with South Korea and stated that they would view any sanctions as an act of war. The North Koreans threatened to withdraw from the NPT. Contingency plans for an air strike against North Korean facilities were even drawn up by the Clinton administration in case the situation worsened.

In early 1994, former President Carter made two trips to North Korea (not sanctioned by the US government at the time) to help diffuse the situation, which led the US and North Korea to negotiate what was called the 'Agreed Framework'.

In return for shipments of heavy oil to fuel its commercial power plants, the DPRK agreed to shut down all its nuclear facilities and remain within the NPT. The US, through a South Korean company, then began constructing two light water reactors to be used to create electricity for the DPRK's commercial use. After the reactors had been built, but before the reactor fuel was in place, the North Koreans would then transfer the 8000 fuel rods from the 5 MW(e) reactor to the IAEA through a third party[5].

The process dragged on and after 2002 President Bush's axis-of-evil speech, the North Koreans admitted to a clandestine uranium enrichment program [11] and both the US and the DPRK stated that the Agreed Framework was invalid. The IAEA inspectors, still in the DPRK, were asked to leave the country and in 2003 North Koreans officially withdrew from the NPT[6]. Although the Agreed Framework was deemed a failure, it did halt their program for eight years.

The Bush Administration began to engage with North Korea through the mechanism of the 'Six Party Talks'. They believed that all countries with a security interest in the region (North and South Korea, China, Russia, Japan, and the US) should meet to discuss a resolution to the matter, and attempt to end North Korea's nuclear program. The North Koreans were not happy with this arrangement, and the early meetings were not fruitful. After many rounds of discussions, all the talks have ended.

Even though strict sanctions are in place, the DPRK has continued to divert its economic resources toward its nuclear program, taking provocative actions whenever they imagine there is a threat, they feel neglected by the international community, or they just want to impress their starving population. They have conducted four nuclear weapons tests, and several missile tests launches, essentially thumbing their noses at international sanctions.

4.8.5 Iran

Iran was one of the original signatories of the NPT in 1968. Under the Shah (Mohammad Reza), Iran began its program in the 1970s. They eventually wanted 23 commercial nuclear reactors from the US through the Atoms for Peace program of President Eisenhower. However, the first contract signed was with a German company to build two reactors at Bushehr. The Iranians purchased uranium from South Africa, and experimented with enrichment and reprocessing technologies.

---

[5] As part of an interim pact with the US and the IAEA, the North Koreans agreed to remove all 8000 fuel rods from their 5 MW(e) reactor and place them in a storage pond. The North Koreans removed the fuel rods without IAEA supervision. The fuel rods were shuffled so that there was no way to tell where in the reactor the rods came from. Had the IAEA been able to supervise the removal of the rods and carefully mark their locations within the reactor on them, the IAEA might have been able to determine how much Plutonium was produced during the reactor's operation.

[6] An interesting note is that although the DPRK withdrew from the NPT, NPT parties do not recognize the withdrawal as legitimate and still view the DPRK as being in violation of their NPT commitments.

In 1979, when the Islamic Revolution overthrew the Shah and put the Ayatollah Khomeini into power, interest waned in nuclear energy and the program was halted. During the Iran–Iraq War in the 1980s, the Bushehr reactor was bombed six different times by the Iraqis. After the costly war, and spurred on by concerns over Iraq's program, the Iranians restarted their reactor program in 1989.

The then USSR, against the objections of the US, agreed to rebuild the Bushehr reactor in 1990. Because of US objections and pressure, Spain, Germany, and China all agreed to end possible agreements on nuclear cooperation with Iran. US concern over nuclear cooperation was based on an assessment that Iran was interested in building a weapon and not peaceful nuclear technology. Although Iran was at that time a good-faith member of the NPT and entitled to access nuclear technology, the US was concerned that once Iran had access, it would drop out of the NPT and develop a weapons program. This so-called loophole in the NPT is of major concern to the US and the IAEA.

These concerns were justified when in 2002 Iran announced the existence of enrichment facilities and a heavy water production facility, and the construction of a heavy water reactor (which would be capable, when operating, of producing large amounts of plutonium). The admission of these clandestine efforts, undertaken without the knowledge of the IAEA, came when Iranian dissident groups in the UK published stories and photographs of these facilities. As a result of this disclosure, Russia amended its agreement with Iran, so that it would maintain control over any spent fuel rods from the Bushehr reactor, meaning Iran would be unable to reprocess them. The IAEA inspected the facilities and declared that Iran had failed to satisfy its NPT obligations. It gave Iran until the end of October 2003 to divulge the extent of its programs.

Iran agreed to temporarily halt its enrichment program in 1983 in order to address IAEA concerns. There were numerous negotiations with the European Union (EU) about nuclear cooperation and assistance if Iran would agree to halt its enrichment program. However, by 2005 Iran had restarted the program. Following this, there were even more rounds of negotiations, sanctions from the EU, the US and the UN, and IAEA inspections, but Iran remained determined to develop an enrichment capability.

The sanctions were beginning to have a significant effect on the Iranian economy and in 2013, with the election of Hassan Rouhani as President of Iran, a more moderate leader, talks began in earnest with Iran and the so-called P5+1[7]. After several rounds of negotiations and missed deadlines, an interim agreement, called the 'Joint Plan of Action', was signed on 24 November 2013. It consisted of a short-term freeze of portions of Iran's nuclear program in exchange for decreased economic sanctions on Iran, as the countries worked towards a long-term agreement. The IAEA was allowed to conduct more intrusive inspections and the agreement was formally activated on 20 January 2014.

---

[7] The P5 are the five permanent members of the UN Security Council: China, France, the United Kingdom, Russia, and the United States. Germany also took part in the talks and is the '+1'.

Finally, after almost two more years of intense negotiations, a final agreement, called the 'Joint Comprehensive Plan of Action' (JCPOA) was signed between Iran, the P5+1, and the EU, which would significantly limit Iran's ability to clandestinely develop nuclear weapons.

There are those who are critical of the JCPOA, but there appear to be very few alternatives other than continued sanctions or military action. Under the strict sanctions, Iran was able to install over 20 000 centrifuges, including over 1000 second generation, more efficient, machines, and assemble a stockpile of more than 10 000 kg of low-enriched uranium (enriched to 3%–5% ^{235}U) and 300 kg of uranium enriched to 20% ^{235}U. They had begun constructing a heavy water reactor that could produce plutonium and had constructed a large underground facility to enrich uranium. Even though the sanctions affected the Iranian economy, they had little effect on Iran's nuclear program. The following is a summary of the JCPOA.

Iran's current stockpile of low-enriched uranium will be reduced from 10 000 kg to 300 kg for a 15-year period and Iran will only be able to enrich uranium to 3.67% for its civilian nuclear power and research. Any enriched uranium in excess of 300 kg will be disposed of, most likely by selling it on the international market for nuclear reactor fuel.

Iran will place over two-thirds of its centrifuges in storage under IAEA control and only be allowed to use 5060 first-generation centrifuges to enrich uranium for the next 10 years. Iran will not build any new uranium-enrichment facilities for 15 years. Iran may continue research and development work on enrichment, but that work must take place under strict controls and only in one designated location. The large underground enrichment facility will be prohibited from enriching uranium for 15 years. An international consortium will convert the facility to an international center for nuclear physics and nuclear technology.

A working group consisting of Iran and the P5+1 will modernize and rebuild the heavy water research reactor in such a way as to minimize the production of plutonium. All spent fuel will be sent out of the country. Any excess heavy water, not needed for the redesigned reactor, will be sold on the international market. Iran will not reprocess spent fuel or engage in reprocessing research for the next 15 years, nor will they be allowed to build any new heavy reactors during that time.

Finally, Iran will implement Additional Protocol agreement that it signed many years ago with the IAEA but never implemented fully. This will allow IAEA inspectors access to any facility that might be related to nuclear activities or the storage of nuclear-related materials. The IAEA is also responsible for monitoring all of the actions that Iran must take over this 15-year period and beyond.

4.9 Lessons learned

Nuclear nonproliferation efforts are extremely difficult. As can be seen from the previous examples, sanctions and military actions may not be enough to deter a country obsessed with developing a nuclear weapons program. The international community needs to be vigilant in confronting those countries that are not complying with their NPT obligations. There also need to be more effective technical and other means to detect such behavior. Finally, more needs to be done to address international and regional stability, to reduce a county's incentive to develop nuclear weapons in the first place.

References

[1] US Congress 1961 *Final Report of the Committee on Commerce, United States Senate, Part III: The Joint Appearances of Senator John F Kennedy and Vice President Richard M Nixon and Other 1960 Campaign Presentations (87th Congress, 1st Session, Senate Report no. 994, Part 3)* (Washington, DC: US Government Printing Office)

[2] Crail P 2009 North Korean nuclear test prompts global rebuke *Arms Control Today* **39** 27

[3] www.ctbto.org/press-centre/press-releases/2013/ctbto-detects-radioactivity-consistent-with-12-february-announced-north-korean-nuclear-test

[4] Gottemoeller R (Assistant-secretary, Bureau of Arms Control, Verification and Compliance) 2012 *Opening Statement at the Conference on Disarmament (Geneva, Switzerland, 24 January 2012)*

[5] www.armscontrol.org/aca/npt13steps.asp

[6] NPT 2005 Recommendations submitted by New Zealand on behalf of Brazil, Egypt, Ireland, Mexico, South Africa and Sweden as members of the New Agenda Coalition *Working paper on Nuclear Disarmament for Main Committee I (4 May 2005)* NPT/CONF.2005/WP.27

[7] 2015 *Draft Final Document (21 May 2015)* vol I, part 1 NPT/CONF.2015/R.3

[8] www.iaea.org/worldatom/About/Profile/mission.html

[9] IAEA 2014 *IAEA 2014 Annual Report* www.iaea.org

[10] www.whitehouse.gov/news/releases/2002/01/20020129-11.html

[11] US Department of State 2002 Press release (16 October 2002, Washington, DC)

IOP Concise Physics

The Midlife Crisis of the Nuclear Nonproliferation Treaty

Peter Pella

Chapter 5

The NPT in crisis

Even though the NPT has, for the most part, been successful in its nonproliferation efforts, it has come up short in a number of situations. Most notable was the failure to detect Iraq's nuclear weapons program while it was a member of the NPT. Today the three pillars of the NPT, safeguards inspections, technical cooperation, and disarmament, are all facing difficult challenges that must be addressed in order to ensure that the NPT remains relevant.

The NPT enter into force in 1970 and since then there have been nine review conferences. The parties were only able to reach consensus in 1970, 1975, 1985, and 2010. Even though the parties agreed to extend the NPT indefinitely in 1995, they were still unable to reach consensus on a final document. Therefore the failure to reach consensus on a final document in 2015 is nothing new. However, some of the issues raised during that conference point to a treaty that is suffering a midlife crisis.

5.1 Safeguards

The responsibilities of the IAEA are growing rapidly, particularly in terms of the complexity and number of safeguards inspections. Because of severe budgetary constraints, the IAEA is continually trying to do more with less money and fewer personnel. The burden of the additional protocols is constantly growing as more states conclude their agreements with the IAEA. In addition a large reprocessing facility in Japan that may be going online soon will require increased monitoring efforts. Finally, the new agreement between Iran and the P5+1 (JCPOA) will require a significant increase in workload, personnel, and money.

While the IAEA's responsibilities are growing, its funding has stagnated. The IAEA's Board of Governors, made up of 35 representatives from its member states, has put in place a policy of 'zero real growth', which has prevented its budget from growing commensurately with the agency's increasing workload. Instead of raising assessed contributions from its member states, thus providing guaranteed funding, IAEA Director General Yukiya Amano has had to appeal for voluntary contributions to carry out the agency's vital tasks. The IAEA's verification budget for 2015 showed only a minimal increase, from approximately 140 million to approximately 147 million USD [1].

In 2013, the cost for monitoring and conducting safeguards inspections in Iran was approximately 14 million USD. In 2014, the IAEA and Iran signed an interim agreement that was to pave the way for the JCPOA. The agreement required increased monitoring, which cost an additional 1 million USD a month [1]. There was a staff of 50 associated with the interim agreement, while a staff of 130–150 will be needed to monitor the JCPOA. It is unclear where the additional personnel will come from.

In his address to the Board of Governors in August 2015, Director General Yukiya Amano stated that implementation of the JCPOA would require about 9.2 million euros (over 10 million USD) per year. He stated that he would not be asking for changes to the 2016 budget, which means that all additional funding will have to come from extra-budgetary voluntary contributions. The total expenditure needed from the time the plan was adopted in October 2015 to the time it is implemented, hopefully early in 2017, will be 960 000 euros (over 1 million USD) per month. The extra-budgetary funds are already almost depleted [2].

The larger question is whether or not the IAEA can supply the personnel and obtain the extra money to fully implement the JCPOA and continue its other safeguard requirements. According to Director General Amano, 'The Agency has nearly six decades of experience of implementing comprehensive safeguards agreements. We are now doing so in 173 countries. We have been implementing the additional protocol for nearly 20 years. We have top-class technical experts, high-tech equipment and state-of-the art analytical laboratories. In short, we have the experience and expertise to conduct the verification and monitoring work set out in the JCPOA. The combination of comprehensive safeguards agreement and additional protocol, together with the verification and monitoring of Iran's nuclear-related commitments under the JCPOA, represents a very robust verification mechanism in Iran [2].'

Although Director General Amano's optimism is laudable, many in the non-proliferation world have doubts. A number of senior inspectors have retired recently and the number of inspectors with advanced degrees has decreased significantly [3]. The number of inspectors from the developed world has also dropped, causing concern in the United States. A major effort to obtain funds from member countries and recruit qualified candidates will be required.

5.2 Technical cooperation

Article IV.1 of the NPT gives states the right to develop indigenous nuclear technology, including reactors, enrichment, and reprocessing, as long as they comply with their NPT obligations.[1] This is a major part of the grand bargain that (NNWS) get for giving up their option of developing nuclear weapons, and it has also proved contentious at previous review conferences. Several states in the developing world felt

---

[1] 'Nothing in this Treaty shall be interpreted as affecting the inalienable right of all the Parties to the Treaty to develop research, production and use of nuclear energy for peaceful purposes without discrimination and in conformity with articles I and II of this Treaty.'

that there were severe restrictions placed on their ability to acquire nuclear technology, even though they had not been found to be in violation of their NPT obligations by the IAEA. These restrictions mostly came from the US and its diplomatic pressure on other nations over concerns about real intentions and the so-called loophole.

Concern about the loophole is a growing problem with some suppliers of nuclear technology and other NPT parties. It is particularly relevant today because of concern about Iran and whether or not the JCPOA will work. According to Article X.1[2], a country has the right to withdraw from the treaty if the country determines that the current situation has 'jeopardized the supreme interests of its country'. Hypothetically, a country could develop a robust nuclear technology infrastructure, including enrichment and reprocessing, with assistance from the IAEA and other NPT parties, and then leave the treaty with no more inspections or restriction on its programs. Since producing the fuel for a nuclear weapon is the most difficult stage in the development of nuclear weapons, the country would be well on its way to becoming a nuclear-weapons state.

Currently, the only options left to the other NPT parties would be sanctions, including UN sanctions at the request of the IAEA, or military action. Certainly sanctions have not deterred North Korea from continuing its nuclear weapons program. A determined country would certainly not be deterred by sanctions alone. Therefore two proposals have been put forward to try to close this serious loophole.

The first proposal is called the 'nuclear fuel bank' and has been championed by the IAEA. One of the reasons a country would want to have its own enrichment capability would be to have a constant supply of fuel for its reactors without the fear of being unable to obtain enriched uranium on the open market because of political pressures or other situations.

The IAEA would maintain a supply of 90 metric tons of low-enriched uranium fuel for nuclear reactors. A country that is a member state of the IAEA could then withdraw the fuel from the fuel bank. The IAEA has established three criteria for a country to obtain fuel from the bank. These are:
1) the supply of fuel to a nuclear power plant is disrupted;
2) the country is unable to secure fuel from the commercial market, state-to-state arrangements, or by any other such means;
3) the country has in place a comprehensive safeguards agreement with the IAEA and is in compliance with this agreement [4].

The country must agree to a supply agreement with the IAEA and pay the full cost to re-stock the fuel stored in the bank. However, the process must not distort the commercial market. Finally, the country is responsible for the spent fuel, and any storage or reprocessing would be under strict IAEA safeguards inspections.

---

[2] 'Each Party shall in exercising its national sovereignty have the right to withdraw from the Treaty if it decides that extraordinary events, related to the subject matter of this Treaty, have jeopardized the supreme interests of its country. It shall give notice of such withdrawal to all other Parties to the Treaty and to the United Nations Security Council three months in advance. Such notice shall include a statement of the extraordinary events it regards as having jeopardized its supreme interests.'

The bank will be fully funded by voluntary contributions and already 150 million USD has been donated, enough to run the bank for 10 years [4]. It will be located at the Ulba Metallurgical Plant (UMP) in Oskemen, a city in northeastern Kazakhstan, and is part of a large global network that includes an additional reserve in Russia, another reserve in the US, and an assurance of supply guarantee from the UK.

Such a fuel supply bank, if it were viewed as reliable, might encourage a country to forego establishing an indigenous enrichment capability and lessen the likelihood of the technology being diverted toward a weapons program. On the other hand, a country might worry that the cost of uranium fuel on the open market could become such that their own enrichment infrastructure costs would seem worthwhile. Establishing a reliable, fair, and practical process will be crucial to the success of the program. Although some in the nonproliferation community are skeptical about its success, the IAEA, the United States, Russia, and the United Kingdom are strongly supportive of it.

A second proposal is that of 'irreversible safeguards agreements'. Once a country signs its safeguards agreement with the IAEA, those agreements would become permanent. So even if the country decided to leave the NPT, it would still have to fulfill its obligations under its safeguards agreement.

Luxembourg submitted a working paper on behalf of the European Union to the 2005 NPT review conference, recommending that states 'affirm as a matter of principle that all nuclear materials, equipment, technologies and facilities, developed for peaceful purposes, of a State Party to the Treaty on the Non-Proliferation of Nuclear Weapons remain, in case of a withdrawal from the Treaty, restricted to peaceful uses only and as a consequence have to remain subject to safeguards' [5]. Germany and France made similar proposals in 2004. The UN Security Council made an attempt to address this issue by passing Resolution 1887 in 2009, which urged states to 'require as a condition of nuclear exports that the recipient State agree that, in the event that it should terminate its IAEA safeguards agreement, safeguards shall continue with respect to any nuclear material and equipment provided prior to such termination, as well as any special nuclear material produced through the use of such material or equipment' [6].

Certainly one would not try to attempt to amend the NPT, as that would require consensus and open up the whole treaty to possible amendments. One way to circumvent this would be to have states include facility-specific agreements in their safeguards agreements with the IAEA. These are called INFCIRC/66 agreements, named after the information circular from the IAEA concerning such arrangements. They do not lapse with withdrawal from the NPT. They can only be terminated if the facility is 'no longer usable for any nuclear activity relevant from the point of view of safeguards or had become practicably irrecoverable' [7].

It certainly might be difficult to convince NNWS to adopt such restrictive measures unilaterally. A good start would be for the NSG to make any future transfer of technology subject to an INFCIR/66 agreement. A country might feel that the penalty for withdrawing from the NPT *and* violating an INFCIR agreement would be too steep a price to pay. Some have even argued that the NWS should lead the way by signing INFCIR/66 agreements on their non-military enrichment and reprocessing facilities. That could be followed by those NNWS with such technologies (perhaps members of the NSG) doing the same. Maybe this would convince other NNWS to follow [8].

It is difficult to block a determined state from developing a nuclear weapons program, even if that state is part of the NPT. North Korea and Iraq (before the first Gulf War) are good examples. How to provide enough roadblocks to the creation of a nuclear weapons program without restricting economic growth and development will be a difficult problem for NPT parties.

5.3 Disarmament

The most contentious of the three pillars is this issue of disarmament. Article VI states that:

> *Each of the Parties to the Treaty undertakes to pursue negotiations in good faith on effective measures relating to cessation of the nuclear arms race at an early date and to nuclear disarmament, and on a Treaty on general and complete disarmament under strict and effective international control.*

The NNWS believe that this is an important part of the benefits that they should receive for giving up a nuclear weapons program. Every review conference has had difficulty in reaching consensus on a final document because the NNWS believe that the five NWS have not lived up to their commitments. There are over 10 000 nuclear warheads in the world today and over 90% of them are in Russia and the United States. In addition, there are thousands more that are in storage ready to be dismantled.

The final document from the 2010 NPT review conference, agreed to by consensus, stated that the NWS made the following commitments:

> a) *[…] rapidly moving towards an overall reduction in the global stockpile of all types of nuclear weapons;*
> b) *Address the question of all nuclear weapons regardless of their type or their location as an integral part of the general nuclear disarmament process;*
> c) *To further diminish the role and significance of nuclear weapons in all military and security concepts, doctrines and policies;*
> d) *Discuss policies that could prevent the use of nuclear weapons and eventually lead to their elimination, lessen the danger of nuclear war and contribute to the non-proliferation and disarmament of nuclear weapons;*
> e) *Consider the legitimate interest of non-nuclear-weapon States in further reducing the operational status of nuclear weapons systems in ways that promote international stability and security;*
> f) *Reduce the risk of accidental use of nuclear weapons; and*
> g) *Further enhance transparency and increase mutual confidence.*
> *The nuclear-weapon States are called upon to report the above undertakings to the Preparatory Committee at 2014. The 2015 Review Conference will take stock and consider the next steps for the full implementation of article VI* [9].

From their perspective, the NWS believe that they have made progress toward disarmament. Since Russia and the United States have the vast majority of nuclear weapons, the focus is generally on them. In 1970 the United States had a stockpile of over 26 000 nuclear weapons that has been reduced to fewer than 5000. Between 1994 and 2013 the United States dismantled almost 10 000 nuclear warheads. Under the New START treaty, which will take effect in 2018, the United States and Russia will have only 1550 deployed warheads [10]. Russia has made similarly drastic cuts.

There was also the 'Megatons to Megawatts' program that ended in December 2013. The United States and Russia agreed to a deal to remove a Russian stockpile of 500 metric tons of highly enriched uranium (95% ^{235}U) that was taken from dismantled Russian warheads. The Russians first 'down-blended' it, which means it was mixed with natural uranium so that it contained 3–5% ^{235}U, and then sold it to the United States as fuel for its commercial nuclear reactors. About 10% of the electrical energy used in the Unites States today comes from former Russian warheads.

On 5 April 2009, US President Barack Obama, speaking in Czechoslovakia, announced his commitment to a massive reduction in nuclear arms. Here is an excerpt from the nuclear-weapons portion of his speech, as displayed on the White House's website [11]:

So today, I state clearly and with conviction America's commitment to seek the peace and security of a world without nuclear weapons. I'm not naive. This goal will not be reached quickly—perhaps not in my lifetime. It will take patience and persistence. But now we, too, must ignore the voices who tell us that the world cannot change. We have to insist, 'Yes, we can.'

Following Obama's speech in April, the G8 (Canada, France, Germany, Italy, Japan, Russia, the United Kingdom, and the United States) reaffirmed these sentiments. In particular, they stated (as published on the White House website [11] and released by the White House's press secretary 8 July 2009) that the G8 'Underscores the commitment to create the conditions for a world without nuclear weapons and welcomes the nuclear disarmament measures taken by nuclear weapons states'.

The case for the United States' commitment to Article VI was also strengthened in the release of the 2010 'Nuclear Posture Review', which described US nuclear weapons policy. It called for a reduction to the number of nuclear weapons and their role in US national security.

The NNWS do not believe that that is enough. They tend to judge according to the progress that has been made since the last review conference. The New START treaty was signed in 2010 and there has been little progress beyond that. Because of strained relations between the United States and Russia over Ukrainian and Syrian issues, Russia has not supported the US initiative to decrease the nuclear arsenals of the two countries to 1000 warheads each. (In fact, the US Congress has required that it give its approval before more cuts are made.) Russia is also concerned about US plans to expand its missile defense system to Europe. Even though the US claims it is to counter the threat from Iran and has offered to include Russia in its missile defense plans, Russia believes that missile defenses in Europe adversely affect its deterrence capability and are destabilizing.

Due to this frosty climate between the United States and Russia, the numbers of nuclear weapons are unlikely to decrease by the time of the next review conference. Even more discouraging for the NNWS, there are new improvements planned for the delivery systems of the warheads themselves in the coming years. This makes it hard for the NWS to claim that they are working towards reducing the role of nuclear weapons in their nuclear policy.

Instead of building new types of weapons, the United States has for the past 20 years upgraded existing weapons systems to extend their life. The United States has spent 8 billion USD to extend the life of its ICBMs until 2030. The navy will be doing the same for its submarine launched ballistic missile (SLBM). Likewise, the air force has begun doing the same for its cruise missiles and B-2 and B-52 bombers.

Beyond these upgrades to existing weapons, work is underway to design new weapons systems to replace the current ones. The navy is designing a new class of 12 nuclear-missile-carrying submarines and the air force is examining whether to build a mobile ICBM or extend the service life of the existing Minuteman III, while it is also developing a new long-range stealth bomber and a new nuclear-capable tactical fighter–bomber. Production of a new guided 'standoff' nuclear bomb, which would be able to glide toward a target over a great distance, is underway, and the air force is developing a new long-range nuclear cruise missile to replace the current one.

Over the next 10 years the United States plans to spend 355 billion USD on the maintenance and modernization of its nuclear forces, according to the US Congressional Budget Office [12]. This is an increase of 142 billion from the 213 billion USD the Obama administration projected in 2011 [13]. It is projected that the United States will be spending over 1 trillion USD over the next 30 years to modernize and improve its nuclear forces [14].

One weapons development is particular troubling. The United States is modernizing its B61 gravity bomb, so named because it is dropped from an airplane. It has been redesigned to be more accurate and have an adjustable yield. It will also be deployed in NATO countries. Because of its low yield and accuracy, it might be tempting to use it, thinking that there would be very little collateral damage and radioactive fallout. The modernization of this one bomb is expected to cost over 10 billion USD [15].

Even though fiscal constraints may decrease the total amount of money available and require that priorities be set, the cost for modernization is enormous. These programs indicate a substantial commitment to current and future nuclear weapons systems that seems to be at odds with the other statements from the Obama administration concerning reducing the number and role of nuclear weapons. It certainly gives the NNWS support for their argument that the United States is not doing enough to fulfill its Article VI commitment.

Russia is also modernizing its nuclear forces significantly. It plans to replace all its old ICBMs and SLBMs by 2025 and has given it a high priority. Because of the secret nature of data on the Russian economy, it is difficult to obtain firm figures on cost. One source has claimed that Russia plans to spend 46.26 billion rubles (1.4 billion USD) on its nuclear weapons systems, while 29.29 billion rubles was spent in 2015 according to the State Duma Defense Committee's report on the draft federal budget for 2014–16 [16]. This is much less than the US spends, but considering the state of the Russian economy and the budget deficits, it still represents a significant commitment.

The other NWS (except the UK) are all modernizing their nuclear forces. France is in the final phase of modernizing its nuclear forces, which is intended to extend the arsenal into the 2050s. They currently have a stockpile of about 300 nuclear warheads deliverable by aircraft, SLBMs, and cruise missiles. The Chinese have been upgrading their nuclear forces over the last 20 years. They are not only developing new delivery systems, but, unlike the other NWS, they are planning to modestly increase the size of their arsenal from the current number of about 250 warheads.

Among the NWS, the UK has done the most in demonstrating its commitment to Article VI of the NPT. It currently has about 200 warheads on SLBMs. These missiles are on Trident submarines that are leased from the US. They eliminated all their other air- and sea-based weapons in the 1990s. There are now serious discussions within the government about whether or not they should modernize or eliminate their nuclear forces.

Some NNWS, because of frustration at the slow pace of disarmament by the NWS, are looking to other fora to try to exert more pressure. Some suggested placing the issue before the UN General Assembly, the international Court of Justice, or some other ad hoc group. In fact, the UN has just convened an ad hoc group for this purpose. At the last NPT review conference, 108 NNWS endorsed the so-called 'humanitarian initiative' and cited the severe impact any use of nuclear weapons would have on the environment and human populations[3]. This was the product of three meetings, starting in 2013, of the Conference on the Humanitarian Impact of Nuclear Weapons. A number of states, including the New Agenda Coalition, also pushed for language that would establish some sort of legal framework that would require the elimination of nuclear weapons.

Of course the NWS reject this idea out of hand, they are not yet ready to appreciate the value of the humanitarian initiative, and nor are they supportive of a time-bound legal framework for the elimination of nuclear weapons. They also worry about the precedent set by addressing this NPT issue outside of the NPT review process. Finally, one should note that not all NNWS have the same opinion about disarmament. The nuclear umbrella of the NWS protects a number of them, and they would feel insecure if those weapons were totally gone while non-NPT NWS existed. If any of the above-mentioned fora hold a nuclear weapons convention that tries to establish a legal framework for nuclear disarmament, the NWS, their allies, and other moderate NNWS are unlikely to attend. This would certainly exacerbate the disagreement over nuclear weapons disarmament.

---

[3] Formerly the 'Austrian pledge'. See 'Humanitarian pledge': www.bmeia.gv.at/fileadmin/user_upload/Zentrale/Aussenpolitik/Abruestung/HINW14/HINW14_%20Austrian_Pledge.pdf.

5.4 Middle East

Although the Middle East situation is not a formal part of the NPT, it has been a contentious issue in each one of the review conferences. Obviously the major concern to Arab states is that there is one known NWS in the region. Even though the official position of Israel is to neither confirm nor deny their possession of nuclear weapons, it is clear that they have a significant arsenal. Also, since Israel is not a member of the NPT, their nuclear facilities are not subject to safeguards inspections. Egypt has been particularly successful in ensuring that their concerns about the Middle East are discussed fully at every RevCon.

At the 1995 review and extension conference, the formal document announcing the indefinite extension of the NPT contained a resolution on the Middle East. That resolution stated that all countries in the region should join the NPT, all nuclear facilities should be under IAEA safeguards and inspections, and the region should seek to become a nuclear weapons and WMD free zone [17]. Similar language also appeared in the final document of the 2000 review conference.

The final document of 2010 promised that there would be a conference in 2012, organized by the UN Secretary-General, involving all countries in the Middle East for the purpose of establishing a WMD free zone in the Middle East. The conference never happened. The US urged postponement, because it felt conditions in the Middle East were not conducive for a meaningful discussion. Since then, much preparatory work was done between 2012 and 2014 in the form of a series of consultations with all the countries in the region by Jaakko Laajava, the Finnish diplomat who has been facilitating the preparations for the planned conference.

However, due to Egypt's and other Arab states' frustration with the slow pace of progress, the final document of the 2015 review conference called for the conference to take place in March of the next year and did not require all countries in the region to be present. The US and the UK could not support such a statement, since it would violate long-standing US policy on the establishment of weapons free zones. According to Rose Gottemoeller, Under Secretary for Arms Control and International Security:

> *Unfortunately, the language related to the convening of a regional conference to discuss issues relevant to the establishment of a Middle East zone free of all weapons of mass destruction and their delivery systems is incompatible with our long-standing policies. We have long supported regional nuclear weapons-free zones, as these zones, when properly crafted and fully implemented, can contribute to international peace, security and stability. We have also stressed that the initiative for the creation of such zones should emanate from the regions themselves, and under a process freely arrived at and with the full mutual consent of all the states in the region* [18].

Clearly the Middle East will continue to be an issue as long as Israel maintains its nuclear arsenal and states in the region cannot agree on a verifiable WMD free zone. The current situations in Syria, Libya, Iraq, and other places will certainly overshadow any effort by the international community to focus on a WMD free zone at this time. Using the NPT review process to force a solution to the Middle East peace process is counterproductive and hurts the global nuclear nonproliferation effort.

5.5 Security assurances

Although they are not mentioned directly in the NPT, security assurances are seen as an important benefit to those NNWS party to the NPT. Security assurances have been discussed at every NPT review conference and remain a problem for many NNWS. Security assurances are pledges by the NWS to the NNWS. There are two kinds of security assurances; one is called a positive security assurance and the other is called a negative security assurance. The former is one in which the NWS, through action by the UN Security Council, pledge to come to the aid of a NNWS that is either attacked by or threatened with nuclear weapons. The latter is one in which the NWS pledge not to use nuclear weapons against a NNWS.

The positive security assurances made by the US (and other NWS) are quite clear:

The United States and the other Nuclear Weapon States have also offered 'positive security assurances' to NPT non-nuclear weapon States. In 1995 the Security Council adopted resolution S/RES/984 taking note of the unilateral assurances given by the Nuclear Weapon States to seek action by the Security Council to provide assistance, according to the relevant provision in the Charter of the United Nations, in the event that a country is threatened or attacked with nuclear weapons [19].

The US position on its negative security policy was first stated in 1978 and reaffirmed in 1995:

The United States reaffirms that it will not use nuclear weapons against non-nuclear-weapon States Parties to the Treaty on the Non-Proliferation of Nuclear Weapons except in the case of an invasion or any other attack on the United States, its territories, its armed forces or other troops, its allies, or on a State towards which it has a security commitment, carried out or sustained by such a non-nuclear-weapon State in association or alliance with a nuclear-weapon State [20].

This was later clarified in the 2010 Nuclear Posture Review (NPR), which stated:

To that end, the United States is now prepared to strengthen its long-standing 'negative security assurance' by declaring that the United States will not use or threaten to use nuclear weapons against non-nuclear weapons states that are party to the Nuclear NonProliferation Treaty (NPT) and in compliance with their nuclear non-proliferation obligations [21, p 15].

Previous nuclear posture reviews have stated otherwise. The first NPR, required by Congress, was issued during the Clinton administration and left the door open for the use of nuclear weapons in response to threats from other WMDs such as chemical or biological weapons. The next NPR, issued during the George W Bush administration, clearly stated that nuclear weapons could be used against threats from WMD.

The NPR does open the door a bit concerning the use of nuclear weapons by stating:

Given the catastrophic potential of biological weapons and the rapid pace of biotechnology development, the United States reserves the right to make any adjustment in the assurance that may be warranted by the evolution and proliferation of the biological weapons threat and US capacities to counter that threat [21, p 16].

The 2010 NPR also states that the United States

will not use or threaten to use nuclear weapons against non-nuclear weapons states that are party to the Nuclear Non-Proliferation Treaty and in compliance with their nuclear non-proliferation obligations [21, p 15].

But that has not always been the case. Just before the first Gulf War in 1991, former Secretary of State James Baker told Tariq Aziz, Iraq's foreign minister, that if 'you use chemical or biological weapons against US forces, the American people will demand vengeance and we have the means to exact it'. Baker said that 'it is entirely possible and even likely, in my opinion, that Iraq did not use its chemical weapons against our forces because of that warning. Of course, that warning was broad enough to include the use of all types of weapons that American possessed' [22].

Additionally, in April 1996, when discussing the possibility of a chemical weapons facility in Libya, then Secretary of Defense William Perry stated that 'if some nation were to attack the United States with chemical weapons, then they would have to fear the consequences of a response from any weapon in our inventory'. Perry did clarify that a bit by saying, 'in every situation that I have seen so far, nuclear weapons would not be required for response' [23].

Finally, the 2010 NPR commits the US to continually improve the capabilities of its conventional forces 'with the objective of making deterrence of nuclear attack on the United States or its allies and partners the sole purpose of US nuclear weapons' [21, p 17]. This of course implies that other options for the use of nuclear weapons are on the table until that time.

Such ambiguities and actions are at odds with public statements on negative security assurances and are worrisome to the NNWS of the NPT. They would prefer a legally binding treaty containing strong, unambiguous language, signed by all five NWS. As stated in the draft final document of the 2015 NPT review conference:

the Conference urges the Conference on Disarmament, within the context of an agreed, comprehensive and balanced programme of work, to immediately begin discussion of effective international arrangements to assure non-nuclear-weapon States against the use or threat of use of nuclear weapons, to discuss substantively, without limitation, with a view to elaborating recommendations dealing with all aspects of this issue, not excluding an internationally legally binding instrument [24].

References

[1] IAEA 2014 *IAEA 2014 Annual Report* www.iaea.org
[2] www.iaea.org/newcenter/statements/introductory-statement-board-governors-64 august 25 2015
[3] Private communication with members of the US delegation to the IAEA 2010
[4] www.iaea.org/OurWork/ST/NE/NEFW/Assurance-ofsupply/documents/Factsheet_LEU_Bank.pdf
[5] 2005 Withdrawal from the Treaty on the Non-Proliferation of Nuclear Weapons, European Union common approach: working paper Submitted by Luxembourg on behalf of the European Union *Review Conference of the Parties to the Treaty on the Non-Proliferation of Nuclear Weapons* (*10 May 2005*) NPT/CONF.2005/WP.32
[6] UN Security Council 2009 S/RES/1887 (24 September 2009)
[7] IAEA 1968 The agency's safeguards system INFCIRC/66/Rev.2 (16 September 1968)
[8] Goldschmidt P 2015 Securing irreversible IAEA safeguards to close the next NPT loophole *Arms Control Today* **45** 15
[9] www.un.org/ga/search/view_doc.asp?symbol=NPT/CONF.2010/50%20(VOL.I)
[10] Buck C (Deputy Chief of Mission, delegation to the Conference on Disarmament Permanent Mission) 2014 Remarks *Third Meeting of the Preparatory Committee for the 2015 Nuclear Non-Proliferation Treaty Review Conference, United Nations (New York, 2 May 2014)*
[11] www.whitehouse.gov
[12] US Congressional Budget Office 2013 Projected costs of US Nuclear forces, 2014 to 2023 www.cbo.gov/sites/default/files/cbofiles/attachments/12-19-2013-NuclearForces.pdf
[13] Miller J 2011 Statement before the Senate Committee on Armed Services Subcommittee on Strategic Forces (4 May 2011) p 5
[14] Wolfsthal J B, Lewis J and Quint M 2014 The trillion dollar nuclear triad: US strategic nuclear modernization over the next thirty years *James Martin Center for Nonproliferation Studies* (7 January 2014) http://cns.miis.edu/trillion_dollar_nuclear_triad/
[15] www.nti.org/gsn/article/defense-spending-bill-requires-report-cost-nukes-europe/
[16] 2013 Russia to up nuclear weapon spending 50% by 2016 *RIA-Novosti* (8 October 2013) http://en.ria.ru/military_news/20131008/184004336/Russia-to-Up-Nuclear-Weapons-Spending-50-by-2016.html
[17] www.un.org/disarmament/WMD/Nuclear/1995NPT/pdf/Resolution_MiddleEast.pdf
[18] Gottemoeller R (Under Secretary for Arms Control and International Security) Remarks *United Nations, New York* http://m.state.gov/md242778.htm
[19] Wood R (Special Representative to the Conference on Disarmament) 2015 Remarks *Non-Proliferation Review Conference, United Nations, New York, 6 May 2015)*
[20] www.ppnn.soton.ac.uk/bb2/Bb2secK.pdf
[21] US Department of Defense 2010 *Nuclear Posture Review Report* www.defense.gov/npr/
[22] Baker J 2010 Testimony before the US Senate Foreign Relations Committee (19 May 2010)
[23] www.armscontrol.org/factsheets/negsec
[24] www.reachingcriticalwill.org/images/documents/Disarmament-fora/npt/revcon2015/documents/MCI-21May.pdf

Chapter 6

Conclusion

The NPT faces many significant challenges. First and foremost is the fact that it is very difficult to convince other countries that they do need nuclear weapons for their security when they are still an important part of national security discussions amongst the NWS. The nuclear policies of the Cold War era have not changed since its end. There needs to be a fundamental shift in the way nuclear weapons are viewed.

In January 2007, former Secretaries of State George Shultz and Henry Kissinger, former Secretary of Defense William Perry, and former Senator Sam Nunn called for 'a world free of nuclear weapons'. This was published as an op-ed piece in the *Wall Street Journal*. They suggested that nuclear deterrence is becoming less useful for national security and also argued that the existence of nuclear arsenals poses a greater risk (in the form of proliferation by worrisome states and terrorist organizations) than abolishing nuclear weapons altogether [1]. The NWS must reassess their policies, unclouded by the Cold War mentality.

Considering the vast disagreement over the final document of the 2015 review conference, it is clear that significant work needs to be done and fresh ideas need to be generated to bridge the differing positions. A second unsuccessful review conference would be a significant blow to the NPT and international nonproliferation efforts. However, predicting what might happen in a subsequent review conference has proven to be quite difficult. No one would have believed after the acrimonious 2005 review conference that a final document could be agreed to in 2010.

There are a number of steps that could be taken to reduce the divide between the Nuclear Weapons States and the Non-nuclear Weapons States who are party to the NPT, increasing the likelihood of a successful 2020 review conference. However, there are significant challenges for implementing any of them. In any case, the effort should be made and all NPT parties must understand the importance of the treaty and strive to ensure its future viability.

The world today is certainly an unsettled place, probably more so than at any time since the end of the Cold War. There are major concerns about domestic and international terrorism; there are civil wars in Syria, Libya, and other parts of the Middle East; there are fragile governments in Iraq and Afghanistan; and there are other issues too numerous to mention. It will be difficult to muster the diplomatic support and national resolve to focus on the NPT and nuclear nonproliferation. But assuming, however unrealistically, that countries can focus on NPT issues, the following steps would be helpful.

Obviously there needs to be some progress among the NWS, particularly the United States and Russia, on nuclear disarmament. The first step would be an attempt to improve the relationship between the US and Russia. Although disagreements will remain concerning the Syrian and Ukrainian situations, the major sticking points for future nuclear disarmament for Russia are the US plan to deploy missile defense systems in Europe and NATO expansion. This will require significant diplomatic efforts, including bilateral meetings at all levels between both governments. The major problem for any US plan to reduce the number of nuclear weapons is the contentious nature of the relationship between Congress and the President. It would be very helpful if both the US and Russia publicly agreed to hold consultations on further nuclear weapons reductions.

During the 2020 preparatory committee meetings, the US needs to strongly restate its negative security assurance commitment in an unambiguous way. Improvements in its conventional weapons capabilities need to be emphasized over nuclear modernization, so that the US does not feel the need to counter WMD with nuclear weapons. Stating this in the next nuclear posture review would be helpful. However, there is no way to know what the next US administration's commitment to nuclear arms reduction will be.

There needs to be significant funding increases for the IAEA. Member states need to be lobbied to increase their funding commitments and the more financially able countries need to increase their voluntary contributions. Strong leadership at the IAEA will be crucial. In addition, there needs to be a concerted effort within the developed world, in particular, to recruit highly capable individuals to work as inspectors and in other roles within the IAEA. Increased staffing and funding are essential to ensure that the IAEA is able to efficiently and effectively perform its essential safeguards mission.

An attempt should be made to close the NPT loophole. As mentioned previously, the Nuclear Supplier Group should make any future transfer of technology subject to an INFCIR/66 agreement. Nuclear Weapons States could sign INFCIR/66 agreements in respect of their non-military enrichment and reprocessing facilities. The members of the Nuclear Suppliers Group can then do the same. This might encourage other NNWS to do it. These actions would help to reduce the probability of a NNWS attempting to develop a nuclear weapons program and would encourage the NWS to make further disarmament efforts.

It is doubtful that any solution to the Middle East WMD free zone will happen soon. The region is certainly consumed with significant armed conflicts and uncertainty. The focus of the international community is on terrorism, civil wars, and long simmering tensions between Sunnis and Shiites. The United States is consumed by its war on terror, support for Iraq and Afghanistan, and the JCPOA between the IAEA and Iran. There seems to be little political will to tackle the issue of a Middle East WMD free zone. However, because of the importance it plays during NPT review conferences, modest progress must be made on the issue.

As stated before, the US does not feel that the current situation in the Middle East is conducive to meaningful discussions at this time. The other complicating factor is that Israel is not a member of the NPT and all other states within the region are. As such, Israel has little incentive to see the process move forward at this time. Unless the United States (and the UK and Canada, who also objected to the Middle East statement in the final document draft at the 2015 NPT review conference) lend their support to the process, progress on the issue is unlikely. Even so, the current rounds of consultations between countries in the region should continue. The US could also encourage Israel to at least acknowledge the usefulness of continuing the process.

Other concerns are always raised at review conference and any progress would have a significant impact, but the prospects are dim. The Conference on Disarmament could break the logjam and reach consensus on beginning discussions on a Fissile Material Cut-Off Treaty. The difficulty is that the consensus is blocked by Pakistan, which is also not a member of the NPT. Convincing Pakistan to drop its opposition has not worked for many years.

The US and China should ratify the Comprehensive Test Ban Treaty. If the US ratified it, the pressure on China to follow would be greatly increased. However, the Republican-controlled Senate is unlikely to ratify it unless large political capital is expended by the Obama administration, and its priorities are elsewhere. In addition, many in Congress (and elsewhere) believe that the large modernization programs proposed by the US might require nuclear testing at some point.

Without any progress on either the disarmament or the Middle East front, hope for a successful 2020 review conference is dismal. The gap between the expectations of the NNWS and the actions of the NWS is a large and contentious one. How much progress occurs over the next five years will be crucial to the viability of the NPT. The most important element of this whole crisis is whether or not the NPT parties can work through their differences within the NPT process and see the value of this significant international instrument. It is to be hoped that by the age of 50, the NPT's midlife crisis will have eased.

Reference

[1] Shultz G P, Perry W J, Kissinger H A and Nunn S 2007 A World Free of Nuclear Weapons *Wall Street Journal* Op-Ed, January 4 2007 www.wsj.com/articles/SB116787515251566636

IOP Concise Physics

The Midlife Crisis of the Nuclear Nonproliferation Treaty

Peter Pella

Appendix

The NPT Treaty

The following is the full text of the NPT [1].

Treaty on the non-proliferation of nuclear weapons

Signed at Washington, London, and Moscow July 1, 1968
Ratification advised by US Senate March 13, 1969
Ratified by US President November 24, 1969
US ratification deposited at Washington, London, and Moscow March 5, 1970
Proclaimed by US President March 5, 1970
Entered into force March 5, 1970

The States concluding this Treaty, hereinafter referred to as the 'Parties to the Treaty',

Considering the devastation that would be visited upon all mankind by a nuclear war and the consequent need to make every effort to avert the danger of such a war and to take measures to safeguard the security of peoples,

Believing that the proliferation of nuclear weapons would seriously enhance the danger of nuclear war,

In conformity with resolutions of the United Nations General Assembly calling for the conclusion of an agreement on the prevention of wider dissemination of nuclear weapons,

Undertaking to cooperate in facilitating the application of International Atomic Energy Agency safeguards on peaceful nuclear activities,

Expressing their support for research, development and other efforts to further the application, within the framework of the International Atomic Energy Agency safeguards system, of the principle of safeguarding effectively the flow of source and special fissionable materials by use of instruments and other techniques at certain strategic points,

Affirming the principle that the benefits of peaceful applications of nuclear technology, including any technological by-products which may be derived by nuclear-weapon States from the development of nuclear explosive devices, should be available for peaceful purposes to all Parties of the Treaty, whether nuclear-weapon or non-nuclear weapon States,

Convinced that, in furtherance of this principle, all Parties to the Treaty are entitled to participate in the fullest possible exchange of scientific information for, and to contribute alone or in cooperation with other States to, the further development of the applications of atomic energy for peaceful purposes,

Declaring their intention to achieve at the earliest possible date the cessation of the nuclear arms race and to undertake effective measures in the direction of nuclear disarmament,

Urging the cooperation of all States in the attainment of this objective,

Recalling the determination expressed by the Parties to the 1963 Treaty banning nuclear weapon tests in the atmosphere, in outer space and under water in its Preamble to seek to achieve the discontinuance of all test explosions of nuclear weapons for all time and to continue negotiations to this end,

Desiring to further the easing of international tension and the strengthening of trust between States in order to facilitate the cessation of the manufacture of nuclear weapons, the liquidation of all their existing stockpiles, and the elimination from national arsenals of nuclear weapons and the means of their delivery pursuant to a Treaty on general and complete disarmament under strict and effective international control,

Recalling that, in accordance with the Charter of the United Nations, States must refrain in their international relations from the threat or use of force against the territorial integrity or political independence of any State, or in any other manner inconsistent with the Purposes of the United Nations, and that the establishment and maintenance of international peace and security are to be promoted with the least diversion for armaments of the world's human and economic resources,

Have agreed as follows:

Article I

Each nuclear-weapon State Party to the Treaty undertakes not to transfer to any recipient whatsoever nuclear weapons or other nuclear explosive devices or control over such weapons or explosive devices directly, or indirectly; and not in any way to assist, encourage, or induce any non-nuclear weapon State to manufacture or otherwise acquire nuclear weapons or other nuclear explosive devices, or control over such weapons or explosive devices.

Article II

Each non-nuclear-weapon State Party to the Treaty undertakes not to receive the transfer from any transferor whatsoever of nuclear weapons or other nuclear explosive devices or of control over such weapons or explosive devices directly, or indirectly; not to manufacture or otherwise acquire nuclear weapons or other nuclear explosive devices; and not to seek or receive any assistance in the manufacture of nuclear weapons or other nuclear explosive devices.

Article III

1. Each non-nuclear-weapon State Party to the Treaty undertakes to accept safeguards, as set forth in an agreement to be negotiated and concluded with the International Atomic Energy Agency in accordance with the Statute of the International Atomic Energy Agency and the Agency's safeguards system, for the exclusive purpose of verification of the fulfillment of its obligations assumed under this Treaty with a view to preventing diversion of nuclear energy from peaceful uses to nuclear weapons or other nuclear explosive devices. Procedures for the safeguards required by this article shall be followed with respect to source or special fissionable material whether it is being produced, processed or used in any principal nuclear facility or is outside any such facility. The safeguards required by this article shall be applied to all source or special fissionable material in all peaceful nuclear activities within the territory of such State, under its jurisdiction, or carried out under its control anywhere.
2. Each State Party to the Treaty undertakes not to provide: (a) source or special fissionable material, or (b) equipment or material especially designed or prepared for the processing, use or production of special fissionable material, to any non-nuclear-weapon State for peaceful purposes, unless the source or special fissionable material shall be subject to the safeguards required by this article.
3. The safeguards required by this article shall be implemented in a manner designed to comply with article IV of this Treaty, and to avoid hampering the economic or technological development of the Parties or international cooperation in the field of peaceful nuclear activities, including the international exchange of nuclear material and equipment for the processing, use or production of nuclear material for peaceful purposes in accordance with the provisions of this article and the principle of safeguarding set forth in the Preamble of the Treaty.
4. Non-nuclear-weapon States Party to the Treaty shall conclude agreements with the International Atomic Energy Agency to meet the requirements of this article either individually or together with other States in accordance with the Statute of the International Atomic Energy Agency. Negotiation of such agreements shall commence within 180 days from the original entry into force of this Treaty. For States depositing their instruments of ratification or accession after the 180-day period, negotiation of such agreements shall commence not later than the date of such deposit. Such agreements shall enter into force not later than eighteen months after the date of initiation of negotiations.

Article IV

1. Nothing in this Treaty shall be interpreted as affecting the inalienable right of all the Parties to the Treaty to develop research, production and use of nuclear energy for peaceful purposes without discrimination and in conformity with articles I and II of this Treaty.
2. All the Parties to the Treaty undertake to facilitate, and have the right to participate in, the fullest possible exchange of equipment, materials and scientific and technological information for the peaceful uses of nuclear energy. Parties to the Treaty in a position to do so shall also cooperate in contributing alone or together with other States or international organizations to the further development of the applications of nuclear energy for peaceful purposes, especially in the territories of non-nuclear-weapon States Party to the Treaty, with due consideration for the needs of the developing areas of the world.

Article V

Each party to the Treaty undertakes to take appropriate measures to ensure that, in accordance with this Treaty, under appropriate international observation and through appropriate international procedures, potential benefits from any peaceful applications of nuclear explosions will be made available to non-nuclear-weapon States Party to the Treaty on a nondiscriminatory basis and that the charge to such Parties for the explosive devices used will be as low as possible and exclude any charge for research and development. Non-nuclear-weapon States Party to the Treaty shall be able to obtain such benefits, pursuant to a special international agreement or agreements, through an appropriate international body with adequate representation of non-nuclear-weapon States. Negotiations on this subject shall commence as soon as possible after the Treaty enters into force. Non-nuclear-weapon States Party to the Treaty so desiring may also obtain such benefits pursuant to bilateral agreements.

Article VI

Each of the Parties to the Treaty undertakes to pursue negotiations in good faith on effective measures relating to cessation of the nuclear arms race at an early date and to nuclear disarmament, and on a Treaty on general and complete disarmament under strict and effective international control.

Article VII

Nothing in this Treaty affects the right of any group of States to conclude regional treaties in order to assure the total absence of nuclear weapons in their respective territories.

Article VIII

1. Any Party to the Treaty may propose amendments to this Treaty. The text of any proposed amendment shall be submitted to the Depositary Governments which shall circulate it to all Parties to the Treaty. Thereupon, if requested to do so by one-third or more of the Parties to the Treaty, the Depositary Governments shall convene a conference, to which they shall invite all the Parties to the Treaty, to consider such an amendment.
2. Any amendment to this Treaty must be approved by a majority of the votes of all the Parties to the Treaty, including the votes of all nuclear-weapon States Party to the Treaty and all other Parties which, on the date the amendment is circulated, are members of the Board of Governors of the International Atomic Energy Agency. The amendment shall enter into force for each Party that deposits its instrument of ratification of the amendment upon the deposit of such instruments of ratification by a majority of all the Parties, including the instruments of ratification of all nuclear-weapon States Party to the Treaty and all other Parties which, on the date the amendment is circulated, are members of the Board of Governors of the International Atomic Energy Agency. Thereafter, it shall enter into force for any other Party upon the deposit of its instrument of ratification of the amendment.
3. Five years after the entry into force of this Treaty, a conference of Parties to the Treaty shall be held in Geneva, Switzerland, in order to review the operation of this Treaty with a view to assuring that the purposes of the Preamble and the provisions of the Treaty are being realized. At intervals of five years thereafter, a majority of the Parties to the Treaty may obtain, by submitting a proposal to this effect to the Depositary Governments, the convening of further conferences with the same objective of reviewing the operation of the Treaty.

Article IX

1. This Treaty shall be open to all States for signature. Any State which does not sign the Treaty before its entry into force in accordance with paragraph 3 of this article may accede to it at any time.
2. This Treaty shall be subject to ratification by signatory States. Instruments of ratification and instruments of accession shall be deposited with the Governments of the United States of America, the United Kingdom of Great Britain and Northern Ireland and the Union of Soviet Socialist Republics, which are hereby designated the Depositary Governments.
3. This Treaty shall enter into force after its ratification by the States, the Governments of which are designated Depositaries of the Treaty, and forty other States signatory to this Treaty and the deposit of their instruments of ratification. For the purposes of this Treaty, a nuclear-weapon State is one which has manufactured and exploded a nuclear weapon or other nuclear explosive device prior to January 1, 1967.

4. For States whose instruments of ratification or accession are deposited subsequent to the entry into force of this Treaty, it shall enter into force on the date of the deposit of their instruments of ratification or accession.
5. The Depositary Governments shall promptly inform all signatory and acceding States of the date of each signature, the date of deposit of each instrument of ratification or of accession, the date of the entry into force of this Treaty, and the date of receipt of any requests for convening a conference or other notices.
6. This Treaty shall be registered by the Depositary Governments pursuant to article 102 of the Charter of the United Nations.

Article X

1. Each Party shall in exercising its national sovereignty have the right to withdraw from the Treaty if it decides that extraordinary events, related to the subject matter of this Treaty, have jeopardized the supreme interests of its country. It shall give notice of such withdrawal to all other Parties to the Treaty and to the United Nations Security Council three months in advance. Such notice shall include a statement of the extraordinary events it regards as having jeopardized its supreme interests.
2. Twenty-five years after the entry into force of the Treaty, a conference shall be convened to decide whether the Treaty shall continue in force indefinitely, or shall be extended for an additional fixed period or periods. This decision shall be taken by a majority of the Parties to the Treaty.

Article XI

This Treaty, the English, Russian, French, Spanish and Chinese texts of which are equally authentic, shall be deposited in the archives of the Depositary Governments. Duly certified copies of this Treaty shall be transmitted by the Depositary Governments to the Governments of the signatory and acceding States.

IN WITNESS WHEREOF the undersigned, duly authorized, have signed this Treaty.

DONE in triplicate, at the cities of Washington, London and Moscow, this first day of July one thousand nine hundred sixty-eight.

Reference

[1] http://www.state.gov/t/isn/trty/16281.htm

www.ingramcontent.com/pod-product-compliance
Lightning Source LLC
Chambersburg PA
CBHW081355230426
43667CB00017B/2845